Restore! Biocapacity and Beyond:
Living Within a Finite Biosphere

Nancy G. Kling

Printed in the United States of America by BookBaby

First Printing, 2020

ISBN:978-1-09831-315-9

Cover Image under license from Shutterstock.com

TABLE OF CONTENTS

For John, who has shared my love of exploring the natural world as my best friend, playmate, and partner

INTRODUCTION

This is *not* another book about climate change. Rather, it is an examination of the overarching issue of our collective failure to live within the limitations of the ecological systems and resources provided by our planet Earth. Ecological systems and resources support all life on our planet and include all the materials consumed by living organisms, along with the processes that recover, cleanse, regulate, and recycle those materials. Stated another way, every substance and raw material required for our existence comes from Earth's biosphere, and all depleted and discarded substances and materials are returned to Earth's biosphere as waste to be assimilated and recycled by Earth's ecological systems. These materials and processes are literally our life support systems. They are also the foundation upon which all economic activity and every other human endeavor are based. Consequently, Earth's ability to provide these essential materials and to regulate these critical processes establishes the limit for all human enterprise.

There is an elephant in the room we have been reluctant to acknowledge. Our elephant is defined by our unrelenting and immoderate (i) extraction and consumption of natural resources, (ii) production and consumption of manufactured products and services, (iii) population growth momentum, and (iv) production of waste and pollution, all buttressed by an economic system that rewards the preceding activities by failing to monetize the negative impacts they generate. It is no wonder we have been reluctant to acknowledge our elephant, as discussions of these unsustainable activities can be charged with emotion and political animus. To compound the problem, most of us have but a rudimentary understanding of the earth sciences that explain the functioning of these ecological resources and systems, so we tend to repeat the talking points from the position we have adopted. There is no denying that the underlying science is challenging; still, I am confident that by the time you have finished reading this book, your appreciation of the operation of Earth's essential systems and resources will have deepened. Without a working literacy of the earth sciences describing our ecological systems and a command of the broader sustainability issues, we are vulnerable to misinformation and intentional disinformation. It is our informed, clear, and compelling voices directed at policymakers, along with our enlightened personal decisions, that together have the power to produce change needed to restore equilibrium to our disrupted natural systems.

Regrettably, for an assortment of political, cultural, and financial reasons, those individuals with the most influential voices in our society have historically forsaken many of these overriding sustainability concerns. These influential, but silent, voices tend to belong to individuals with leadership roles in government and some of our most significant corporate organizations. It is difficult for these voices to acknowledge that we are living beyond the capacities of the biologic, geologic, atmospheric, and hydrologic systems of our planet, if they have no intention of initiating remedial action. It is difficult for these voices to acknowledge that increasing levels of production and consumption of manufactured products and services cannot be supported indefinitely by Earth's resources and systems, if they fear that asking us to consume less will hurt their wallets. It is particularly difficult for these voices to acknowledge that we have more people living on our planet than the planet can adequately support, if they fear that cultural and religious concerns that swirl around population discussions could endanger their leadership positions. My indictment notwithstanding, I am relieved to observe the embryonic beginnings of a public conversation about the broad environmental impacts of continued population growth and our current livestock agricultural practices. The United Nations and numerous nonprofit organizations have been trying for decades to take a leadership role in these broader sustainability concerns. At the end of the day, it will become evident that sustainable practices in industry and agriculture are possible without becoming unprofitable. Furthermore, it will become evident that we can have prosperity without increasing production, consumption, and the number of consumers on the planet. The more consequential question is, when will that day occur?

We are disturbing the natural systems and resources of our planet in several fundamental ways: (i) we have disrupted the generally balanced condition of the natural biogeochemical cycles of our biosphere (e.g., water cycle, carbon cycle, oxygen cycle, climate cycle, methane cycle, phosphorus cycle, and nitrogen cycle), (ii) we have degraded many critical natural resources, such as land, water, air, soil, and other resources essential for human life, (iii) we have diminished many of our nonrenewable resources for which we have no replacement, (iv) we are experiencing a global pattern of population growth momentum, giving rise to unsustainable levels of consumption of goods and services, and (v) we have continued to support industrial livestock agriculture that turns out to be one of the most environmentally destructive business sectors. This book will explore each of the above subjects, but will avoid delving any more deeply into the political, cultural and financial reasons behind the absence of these subjects from the public discourse—we will leave that for the political scientists, sociologists, and psychologists.

Humans have been modifying the natural biogeochemical cycles of this planet from the very beginning of our existence as *Homo sapiens* (the species to which all modern humans belong), but the scale of our impacts since the Industrial Revolution has been profound and transformative. Except for a handful of political holdouts, most environmental scientists concur that our current consumption of resources, disruption of the natural biogeochemical cycles, and creation of waste and pollution are not sustainable. The same clock that is running down with regard to climate system disruption is also running down with regard to the capacity of the natural systems and resources of this planet to sustain our population, both current and projected.

Even for those of us with no vested interest in quieting public discourse surrounding these larger sustainability issues, we should not forget how laborious it has been for us as a global society to fully process the message about fossil fuel emissions and climate system disruption. If our collective brains have struggled to fully comprehend what we, our governments, and our policymakers need to do to reduce the harm from fossil fuel emissions, how can we possibly grasp the notion that fossil fuel combustion is only the tip of the sustainability iceberg? Some scientists have hypothesized that the flip side of our hardwired fight-or-flight mechanism is our inability to process the risk of events distant in time and place—that we are hampered by a kind of cognitive nearsightedness for such remote consequences. Perhaps our unsustainable activities constitute so weighty a problem that we can only perceive a small portion of it at any given time and assume that small portion constitutes the crux of the problem. It reminds me of the ancient Indian parable of the five blind men and the elephant, where each man characterizes the entire elephant to be just like the small area of the elephant he discerns with the touch of his hands. I intend to address the most serious of these sustainability issues, provide an accessible explanation of the underlying science, and create an economic framework to enlarge the scope of our public discourse. Let's take a good, hard look at our entire elephant.

We will begin by reviewing some basic principles of ecology in Chapter 1. In addition, we will define and discuss the notions of sustainability, carrying capacity, and biocapacity. In Chapter 2, we will explore the components of some of the generally balanced biogeochemical cycles we have managed to disrupt. We will also examine some of the critical resources we are consuming or damaging faster than they can regenerate or otherwise be replaced. As we examine each ecological cycle and resource, you will see how humankind has impacted the biological, chemical, and geological functioning of each cycle or resource. Chapter 3 will continue this analysis, but will concentrate on the biogeochemical systems that produce substances implicated in creating greenhouse effects. In Chapter 4, we will take a look at population dynamics—past, present, and projected. As indicated earlier, this is a sticky subject; still, we will examine how the composition of specific population groups is changing, and how these demographic changes are impacting the ecological resources and systems that sustain us. In Chapter 5, we will take a look at the widespread and harmful effects of livestock agriculture, a significant producer of methane emissions, nitrous oxide emissions, water pollution, eutrophication, water scarcity, and deforestation. We will explore how these impacts affect the functioning of our biogeochemical cycles and natural resources. Armed with an increased understanding of ecology and the earth sciences, we will turn our attention in Chapter 6 to some of the cutting-edge sustainability technologies. These advanced technologies promise to offset a portion of our environmental degradation, disruption, and depletion, and perhaps even buy us some time to take the actions necessary to prevent us from further exceeding the biocapacity of our planet. In the final chapter, I will tie this all together with a discussion of the relationship between the economy and our ecological resources and systems.

Young people from all over our planet seem to comprehend that we are living with some very sobering environmental conditions, most of which have been created over the past several generations by their parents, grandparents, and great-grandparents. These conditions will affect young people in ways none of us can imagine. I am impressed by their activism and their ability to articulate their concerns. I am part of those past generations, and I hope to advance the discussion and possibly point it in the direction of remedies that can provide the most benefit in the shortest time. I hope that this information is easy for the reader to access and that it provides tools for the reader to make the case that the environmental issues confronting us are much greater than our collective carbon footprints.

Chapter 1

BIOCAPACITY PLANET EARTH

Before we head into biocapacity, we should review a few of the basic principles of ecology and sustainability that we are going to use throughout this book.

ECOLOGY: Ecology is the study of how living things interact with each other and with their physical environments. How living things interact is a function of the other biotic organisms (e.g., living organisms, such as plants, animals, bacteria, and fungi) and the abiotic elements (e.g., non-living elements, such as temperature, light, wind, water, atmosphere, and soil) of their immediate environments. We will look at the roles played by biotic and abiotic factors as we examine each of the ecological systems we have disrupted.

ECOSYSTEM: This is a much-used term and warrants a clear definition. An ecosystem is a community of biotic organisms and abiotic elements in a defined area that are interdependent, functioning together as an ecological unit. The defined area can be as small as a cup of soil filled with bacteria, fungi, and nematodes, or as large as an entire tropical rainforest or our entire planet. Ecosystems can be on, in or near water (aquatic) or land (terrestrial).

BIOSPHERE: The biosphere is the aggregate of all of Earth's ecosystems, as found in the atmosphere (air), the hydrosphere (water), and the lithosphere (the crust and the upper portion of the mantle).

MATTER: Since you first started school, you have heard that matter can be neither created nor destroyed. But have you ever stopped to think about what this means in the context of our ecological systems? It means that nearly all of the atoms in the molecules that form the components of our biosphere have been around since our planet was formed. It means our ecological and geological systems keep on recycling, cleansing, decomposing, and reconsolidating all of the materials needed to sustain plant, animal, and human life. The carbon and hydrogen atoms you pumped into your car yesterday have been used over and over many times, in many different forms and for many different purposes.

ENERGY: Most of the energy driving planet Earth comes from the sun. This radiant energy is essential for all the chemical reactions that sustain life. Our sun is nothing if not a massive fusion reactor whereby hydrogen atoms under extreme heat and pressure in the core of our sun undergo fusion reactions that release energy. That energy travels through space, into and through Earth's atmosphere, to its destination at the surface of our planet. At the surface of our planet, that solar energy provides all living organisms, either directly or indirectly, with the energy required to drive the biological and chemical reactions that sustain life.

The following is a very simplified example of how energy flows through our biosphere:

- Deep inside the core of our sun and in a series of steps, the protons in the nuclei of hydrogen atoms fuse to become one helium atom, releasing energy in the form of photons.

- A stream of photons strikes the surface of Earth.

- Plants absorb the photons, along with carbon dioxide from the atmosphere and water from the soil.

- Plants use these ingredients to produce glucose for their nourishment and release oxygen as a waste product. This chemical reaction is called photosynthesis. The chemical formula for photosynthesis describes the process more elegantly than words:
  $solar\ energy + 6CO_2 + 6H_2O \longrightarrow C_6H_{12}O_6 + 6O_2$.

- Animals eat the plants, and during the process of digestion, the chemical bonds that hold together the atoms that make up the glucose molecules are broken, releasing energy for the animals to use for their growth and biological processes.

- Larger animals eat smaller animals, and the same process takes place, albeit with some loss of energy.

- When animals die, their remains are consumed by bacteria, fungi, and other decomposers in the soil. The decomposers transfer the remaining energy in the tissues of dead animals to the soil via the chemical compounds in their excretions.

Another source of energy that frequently gets short-changed in discussions about energy is the energy produced by chemosynthetic organisms that live in and around deep-sea hydrothermal vents and methane seeps. In a process that is very different from photosynthesis, these chemosynthetic organisms, often in the form of bacteria, use the hydrogen, nitrogen, methane, and other gas emissions from these vents and seeps, along with seawater and dissolved carbon dioxide, to make glucose. The glucose then becomes an energy source for other sea life. This energy is produced entirely in the dark and relies on chemicals rather than photons from the sun.

BIOGEOCHEMICAL CYCLES: These are the processes that recover and recycle all of the elements critical to life on Earth. These essential elements were formed during the creation of our universe 13.5 billion years ago. Scientists believe that explosions of early stars spewed stellar materials into space to form other stars, which in time also exploded. About 4.5 billion years ago, a molecular cloud of stellar gas and dust collapsed from its own gravitational pull to form our solar system's early Sun, with the remaining dust and gases spinning around it in a solar nebula. According to the core accretion model, gravity started pulling all this swirling gas, dust, and particulate together to form, among other things, the planet we call home. The materials that came together to form Earth have been cycling through Earth's hydrosphere, lithosphere, and atmosphere for the last few billion years. Some of the biogeochemical cycles we will explore are the water, carbon, oxygen, nitrogen, methane, and phosphorus cycles.

ECOSYSTEM SERVICES: Ecosystem services include all of the benefits provided by the natural environment and its diverse ecosystems. They directly or indirectly support all life on our planet. These services have been widely recognized by the international community through the United Nations and are generally grouped into four categories: (i) provisioning services such as food, fresh water, fuel, and minerals, (ii) regulatory services such as climate regulation, water purification, and pollination, (iii) supporting services such as photosynthesis, soil formation, and nutrient cycling and (iv) cultural services such as the recreational and aesthetic benefits of the natural environment.

BIODIVERSITY: A healthy and functioning ecosystem has the resilience to survive most of what Mother Nature throws at it. In a healthy ecosystem, plants and animals can adapt to changes and challenges. One of the most critical factors that can keep an ecosystem healthy is biodiversity. A biodiverse ecosystem offers animals many alternatives for obtaining food, water, and shelter, offers a more complex and diverse food web, and provides a variety of pathways for critical elements to cycle and recycle through the ecosystem.

PLANTS: It is important to remember that plants, algae, and certain bacteria are the *only* organisms on Earth that can capture solar energy directly from the sun. Their ability to convert solar energy to chemical energy and make that energy available to all other living things is remarkable. Because of this unique ability, plants are called producers, and the rest of us are consumers. Plants are also one of the primary sources of oxygen required for respiration by animal life. Plants require certain nutrients to thrive, such as carbon, hydrogen, oxygen, nitrogen, potassium, and phosphorus. When we disrupt biogeochemical cycles that continuously provide us with these nutrients, we are making it more difficult for plant life to survive. If our plants don't survive and thrive…well, that is a gamechanger, probably for all but extreme microbial life living deep in Earth's crust and oceans, such as the chemosynthetic organisms we discussed in the section on energy.

ENVIRONMENTAL SUSTAINABILITY: If everything humans require for survival comes from the natural environment, then what exactly is environmental sustainability? My preferred definition comes from Herman Daly, a pioneering American ecologist, economist, and a 2020 nominee for the Nobel Peace Prize for sustainable development. He divides the environmental universe into three useful categories: (i) renewable resources, (ii) nonrenewable resources, and (iii) waste products and pollution. The following is my restatement of Daly's three principles of sustainability.

- A sustainable use of renewable resources means that our rate of consumption is not greater than the rate at which Earth's biogeochemical cycles can recycle or regenerate them.

- A sustainable rate of discharge of pollution and waste means that pollution and waste should not be discharged faster than the rate at which Earth's biogeochemical cycles can assimilate them, recycle them, or render them harmless.

- A sustainable use of nonrenewable resources means that our rate of consumption should not be greater than the rate at which their renewable substitutes can be developed and deployed.[1]

Environmental sustainability, by definition, allows for consumption and waste generation at rates no greater than that which can be continued indefinitely. If a particular level of consumption and waste cannot be continued indefinitely, then the rate is not sustainable, and demand must be reduced if we are to avoid exceeding the biocapacity of our planet.

BIOCAPACITY: Biocapacity is sustainability's first cousin. It refers to the ability of existing ecosystems to provide the essential natural resources and assimilate and renew the waste produced by human enterprise, indefinitely. When we exceed the biocapacity of an area, it can be said that our activities in that area are unsustainable. Biocapacity is usually expressed in hectares per capita and tells us how many humans can sustainably live in a given area before a biocapacity deficit occurs. The graphic below from Global Footprint Network, a sustainability think tank, is an excellent depiction of our current biocapacity deficit; that is, the extent to which our consumption exceeds the planet's biocapacity, expressed as hectares per capita.

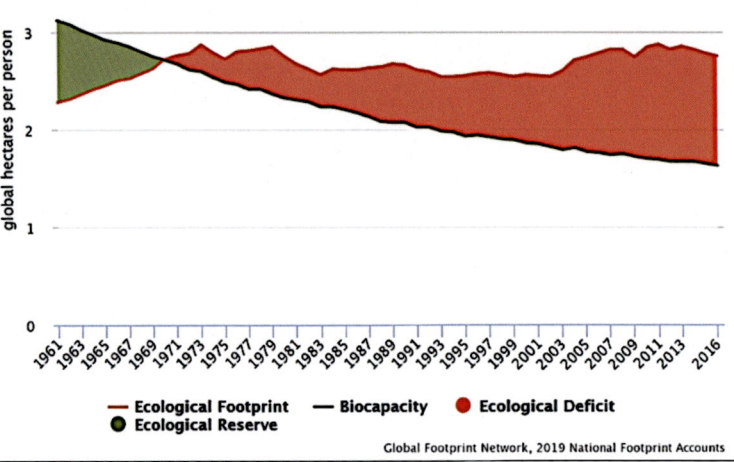

World Ecological Footprint
Credit: Global Footprint Networks as open source material.
http://data.footprintnetwork.org/?_ga=2.60038637.1137230436.1587751622-1994419756.1587229084#/countryTrends?cn=5001&type=BCtot,EFCtot

Global Footprint Network, 2019 National Footprint Accounts

CARRYING CAPACITY: This term is defined in the scientific literature as the number of plants, animals, people, or other living organisms in a designated area that can be supported indefinitely by available natural resources and ecological systems without environmental degradation. In the case of the carrying capacity for humans on planet Earth, carrying capacity is reduced when too many people make demands on existing ecological resources and systems. Carrying capacity is generally expressed as a number of persons, rather than a ratio of hectares per person as with biocapacity.

So, what is the upper limit of the carrying capacity of our planet, and more importantly, is that limit in our rearview mirror or ahead of us? And if it is ahead of us, how far ahead? Since I fully intend to get out of bed tomorrow morning, I have to conduct myself as though the limit is still ahead of us. Every species has its own carrying capacity for its particular territory, but *Homo sapiens* is a complex animal, who consumes, expands his territory, reproduces, and generally interacts with his environment in complex and unpredictable ways. The forces that drive the behavior of *Homo sapiens* go well beyond the biological mandate of his survival and the survival of his species, owing to his enormous brain and attendant complex cognitive skills.

Calculations to determine carrying capacity for *Homo sapiens* require projections and predictions about reproduction, economics, technological advances, accessibility to natural resources, demographic patterns, and a host of other variables. Most estimates in the scientific literature of Earth's current carrying capacity fall in the range of nine to ten billion people. Consequently, ten billion has more or less become the working carrying capacity valuation. Our current global population as of December 2019 is 7.7 billion and is projected by the United Nations to be at 9[+] billion by 2050 and between 11.5 and 16 billion by 2100. The Global Footprint Network, in collaboration with World Wide Fund for Nature, has estimated that if everyone consumed as much as the average European (which is 60% of the average American), then we would need 2.8 Earths right now. Stated another way, if everyone on the planet consumed at the same rate as the average European, our planet could generate adequate resources and assimilate humanity's waste and pollution for only about two to three billion people.[2] Don't even think about doing the calculation to determine Earth's carrying capacity based on the consumption level of the average American; it's not an uplifting number.

A group of intrepid minds from the Stockholm Resilience Centre, an internationally recognized nonprofit research institute dedicated to long-term sustainability and environmental issues, has attempted to define the safety zones within which we humans can live sustainably. For each of Earth's critical ecosystem resources and systems, they identified the proxy that best measures that particular ecosystem resource or system (e.g., concentrations of carbon dioxide as a proxy for climate change). They then took the additional step of quantifying the biocapacity limits for each measure (e.g., parts per million of atmospheric carbon dioxide). Finally, they have demonstrated quantitatively the level at which we are currently operating for each system or resource. I refer you to *Big World Small Planet: Abundance within Planetary Boundaries*[3] and to *Ecological Footprint: Managing our Biocapacity Budget.*[4] As you may well have guessed, we have exceeded the safety zone on several of the parameters. We will discuss this exciting project in further detail in the final chapter.

Whatever the limits of our biocapacity or carrying capacity may be, I hope by the last page of this book, you will agree we cannot continue to consume as though the ecological systems and resources upon which we rely will always be there for us with sufficient quantities and functionality. Dr. Daly reminds us that we cannot continue to consume as though Earth were a business in liquidation.

In the following chapters, we will explore each ecological system and critical resource from the perspective of Dr. Daly's definition of sustainability. Specifically, we will examine how humankind's activities are (i) *damaging and depleting* renewable resources faster than they can regenerate, (ii) *diminishing* nonrenewable resources before renewable substitutes have been developed, and (iii) *disrupting* natural biogeochemical cycles and their ability to assimilate, recycle and regenerate our waste and pollution. You can remember these activities as the *3 Ds*. The *3 Ds* are together impacting the biocapacity and carrying capacity of our planet by reducing the amount of time we have ahead of us before the ecological systems and resources of our planet are so disrupted that life on planet Earth is far less hospitable.

The following is Thomas Malthus's somewhat hyperbolic, but also somewhat accurate, view on carrying capacity, written in 1798. By way of introduction, Malthus was a recognized economist of his time and one of the early scholars broaching the issues of overconsumption and overpopulation.

The power of population is so superior to the power in the Earth to produce subsistence for man, that premature death must in some shape or other visit the human race. The vices of mankind are active and able ministers of depopulation. They are the precursors in the great army of destruction; and often finish the dreadful work themselves. But should they fail in this war of extermination, sickly seasons, epidemics, pestilence, and plague, advance in terrific array, and sweep off their thousands and ten thousands. Should success be still incomplete, gigantic inevitable famine stalks in the rear, and with one mighty blow levels the population with the food of the world. [5]

While one might view this quote as a dark view of humanity's relationship to the planet, it is difficult to argue with the premise that *Homo sapiens* has aggressively diminished and degraded the resources of this planet and, in the process, severely disrupted many of its biogeochemical systems. To be fair, we should cut Malthus a bit of slack, as he could not, in 1798, have envisioned the technological changes in food production that would radically increase the per-acre yield of crops or the massive expansion of global trade that would distribute scarce resources around the world. On the other hand, let's not be too generous with Malthus, as he appears to have just been a trifle too obsessed with man's inability to resist the attractions of the fairer sex, especially when he names this attraction as the fundamental cause of uncontrolled population growth that will ultimately kill us all. "It is a truth, which history I am afraid makes too clear, that some men of the highest mental powers have been addicted not only to a moderate, but even to an immoderate indulgence in the pleasures of sensual love." [6] As I mentioned, his exaggerated comments read somewhat humorously today, yet there is some truth in his comments about population growth.

BOTTOM LINE: BIOCAPACITY: In an ideal world, we would enjoy replacement-level population growth, stabilized at a level capable of being supported by existing ecological resources and systems, indefinitely. Our ecological systems would be generally balanced, closed-loop cycles that recycle and regenerate renewable resources, with no reduction in the availability of those resources. We would not consume more nonrenewable resources than we can replace with renewable alternatives, and the waste and pollution we create would not exceed the ability of the environment to assimilate that waste and pollution. The graphic below is an aspirational depiction of a balanced Earth ecosystem.

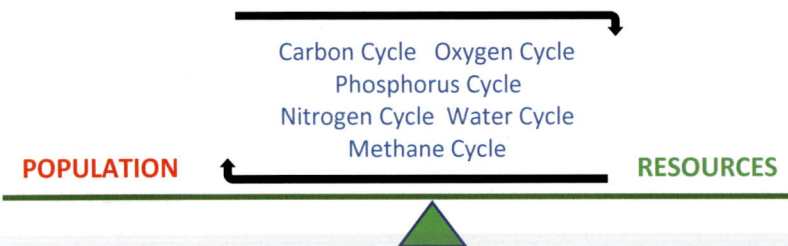

Timeline----------------------------------*THEORETICAL INFINITY* ---------->

Our current trajectory of population growth and resource consumption and degradation has us severely diminishing our reserves of critical resources and having a large share of other resources becoming unusable. This trajectory also has our natural biogeochemical cycles becoming increasingly dysfunctional. Each ecological cycle that is disrupted and each resource that is damaged or diminished shortens the timeline remaining before we reach the point where the planet cannot adequately sustain us.

Chapter 2

CRITICAL THREATS TO BIOCAPACITY

In Chapter 1, we reviewed several related terms: sustainability, carrying capacity, and biocapacity. The distinctions among those terms are less important than the fundamental notion that our planet's ecosystem services provide all the materials needed for us to survive and thrive, along with all the recovery, cleaning, recycling, and assimilating of those materials. In this Chapter 2, we will explore specific ecological systems such as the water cycle, nitrogen cycle, and phosphorus cycle, and natural resources such as soil, arable land, forests, rare earth elements, and nonrenewable metals. In Chapter 3, we will explore the carbon cycle, oxygen cycle, and methane cycle, along with the nitrous oxide portion of the nitrogen cycle and the water vapor portion of the water cycle. As we explore each cycle and resource in chapters 2 and 3, it is important to bear in mind a concept from the Introduction that cannot be repeated enough times: *Every substance and raw material required for our existence comes from Earth's biosphere, and all depleted and discarded substances and materials are returned to Earth's biosphere as waste to be assimilated and recycled by Earth's ecological systems. These materials and processes are literally our life support systems. They are also the foundation upon which all economic activity and every other human*

endeavor are based. Consequently, Earth's ability to provide these essential materials and to regulate these critical processes establishes the limit for all human enterprise.

NITROGEN AND THE NITROGEN CYCLE

<u>REACTIVE NITROGEN</u>: Nitrogen is an essential element for all plant and animal life, although excess amounts can be harmful. Nitrogen gas molecules (N_2) make up 78% of our atmosphere, and nitrogen gas consists of two nitrogen atoms with three strong bonds holding the atoms together. Not many organisms can use nitrogen in this form. As a result, nitrogen gas must be converted to a usable form called *reactive nitrogen*. This conversion is the job of the nitrogen cycle; consequently, our lives are dependent upon a healthy and functioning nitrogen cycle. The following is an abbreviated summary of the steps of the nitrogen cycle.

- *Nitrogen fixation*: Microorganisms in the soil harvest gaseous nitrogen molecules from the atmosphere and chemically convert them to ammonia or ammonium. Similar microorganisms live in nodules on the roots of specific plants, such as legumes and clover, where they also harvest gaseous nitrogen molecules from the atmosphere and convert those molecules to a form that can be absorbed by the plant.

- *Nitrification*: The ammonia and ammonium in the soil are converted to nitrates by specific bacteria in the soil called nitrifying bacteria.

- *Assimilation*: Nitrates and other nitrogen compounds are taken up from the soil by plants through their root systems and assimilated into the plant tissues to make vital cellular products. When animals eat the plants, the animals assimilate the nitrogen compounds to make proteins and other molecules needed for their cell functioning. When an animal eats another animal, it also receives the nitrogen compounds from that animal.

- *Ammonification:* When plants or animals die or when animals excrete waste, the nitrogen compounds in their bodies and waste enter the soil, where they are broken down by other microorganisms known as decomposers. This process creates ammonia and nitrates once again.

- *Denitrification:* The ammonia and nitrates in the soil are then converted once again into gaseous nitrogen by microorganisms.

- *Lightening:* There is another method by which atmospheric nitrogen is converted to a form usable by plants. An electric discharge in the clouds can cause atmospheric nitrogen to react with atmospheric oxygen to form nitric oxide. The nitric oxide then reacts with additional atmospheric oxygen to form nitrogen dioxide, which then reacts with falling rainwater to form nitrates and nitrites, which can be absorbed by the roots of plants.

In sum, the nitrogen cycle converts atmospheric nitrogen gas to reactive nitrogen, which is absorbed and used by plants for growth and survival. Animals and humans obtain these essential nitrogen compounds when they eat the plants. Take a moment to study the graphic of the land-based nitrogen cycle shown on the following page.

Although not depicted in the graphic, the marine nitrogen cycle has the same steps, but different marine microorganisms perform the work. Note also that the graphic distinguishes human-based nitrogen sources from natural sources, an important distinction when we are examining nitrogen.

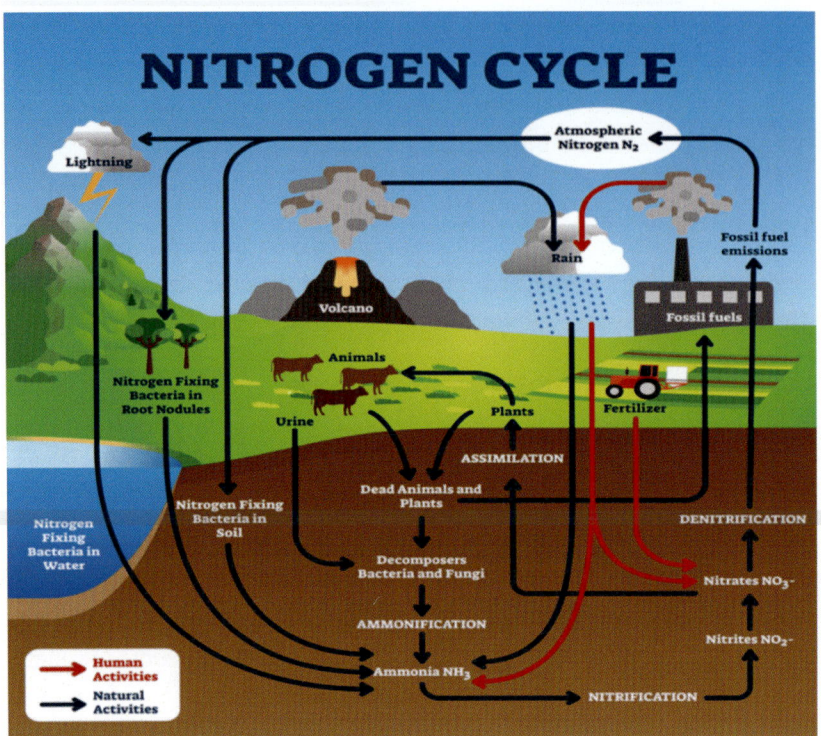

Image under license from Shutterstock.com

Before the advent of fertilizers, farmers planted legumes, clover, and other cover crops capable of fixing nitrogen from their root nodules to restore nitrogen to depleted soils. Applying animal waste or "night soil" from humans was another method farmers used to restore nitrogen to their soil. Another process involved crop rotation, where crops intended for animal grazing were grown alternatively with food crops. The grazing animals provided organic waste that nourished the soil.

In the late 1800s, when demand for food outstripped the yields from these methods, farmers resorted to the importation of bird and bat excrement (guano). Still, food productivity could not keep up with population growth. Because of increasing demand for food crops to feed a growing human population and increasing demand for feed crops to feed a growing livestock population, farmers have resorted to using synthetic nitrogen as fertilizer to enrich their soils and increase their crop yields.

Most synthetic nitrogen is produced using the Haber-Bosch process, which converts free atmospheric nitrogen to ammonia (NH_3) with the use of high pressure and high temperatures. It fundamentally does the work of our decomposers, but it is dependent on large amounts of energy to create high temperatures and pressure. The Haber-Bosch process, developed in 1910, remains the primary method for producing synthetic nitrogen fertilizer. Synthetic nitrogen has materially increased crop yields, and the Haber-Bosch process was a global game-changer. Most researchers concur that commercial fertilizer is responsible for about 40% to 60% of today's global crop yield and that globally we are able to feed twice as many people as we could without the use of commercial fertilizer. Synthetic nitrogen also gave farmers the ability to produce additional grains to feed to livestock and laid the groundwork for the large-scale industrial livestock agriculture we see today. The downside to this increased productivity is that synthetic nitrogen fertilizers are polluting our fresh water supplies, depleting our soils of other nutrients, suffocating marine plants and animals, and releasing greenhouse gases. There can be *too much* of a good thing.

In the US, we are currently introducing into our environment many times the amount of reactive nitrogen than would naturally be available through the nitrogen cycle. Sabrina Shankman, a reporter for InsideClimate News, wrote the following: "The use of fertilizer in the United States has risen more than 200% over the past 60 years, even though the amount of cropland has remained relatively constant. It is estimated that corn crop production would decrease by 40% without added nitrogen. Added nitrogen can take the form of synthetic fertilizers produced by the Haber-Bosch process or organic fertilizers such as manure. About half of the reactive nitrogen applied is taken up by the plants, with the remainder washed away as groundwater or off-gassed as nitrous oxide."[7]

Nitrous oxide is a powerful greenhouse gas we will discuss in the next chapter. When irrigation water or stormwater carries away reactive nitrogen from farms to aquatic and marine ecosystems, the fresh water or salt water receives many times more nitrogen than it can assimilate. This overabundance of reactive nitrogen in the water stimulates excessive growth of naturally-occurring algae, which grow in such quantities that the algae block sunlight and interfere with aquatic or marine plant and animal functions. When the excessive algae blooms die, they sink to the bottom, and bacteria swarm to digest the material. The bacteria usurp much of the oxygen in the water, suffocating aquatic or marine animal and plant life. This process is called eutrophication, and it causes fish kills, loss of plant life, degradation of drinking water sources, loss of biodiversity, human health problems, and losses of billions of dollars per year.[8] With current population projections and increasing consumption of meat products, the demand for food and feed crops will grow, as will the use of reactive nitrogen fertilizer.

BOTTOM LINE: NITROGEN: Historically, reactive nitrogen has been considered a limiting factor, a variable that limited the amount of food that could be produced from our cultivated land. You can better appreciate Malthus's position when you consider that in the 1800s, the amount of food that could be produced was limited by the amount of naturally-occurring reactive nitrogen in the soil. Current levels of food production require many times more nitrogen than is naturally available in our tired, worn-out soils. Consequently, we have become increasingly dependent on synthetic nitrogen fertilizers to grow enough food to satisfy demand. This excessive use of reactive nitrogen is producing significant damage to freshwater supplies and marine environments, as well as emissions of nitrous oxide, a robust greenhouse gas we will cover in the next chapter.

PHOSPHORUS AND THE PHOSPHORUS CYCLE

Interesting factoid: German merchant Hennig Brand discovered the element phosphorus in 1669, having prepared it from numerous buckets of human urine, using evaporation, heat, and condensation. Fortunately, we no longer have to collect buckets of urine to acquire phosphorus for agriculture, because phosphorus is mined from phosphate rock. (Spoiler alert: we will revisit the idea of harvesting phosphorus from urine later—Hennig was actually on to something.)

PHOSPHORUS CYCLE: The graphic on the following page depicts how the phosphorus cycle naturally provides us with this vital element, and the following is a description of each of the cycle steps.

- *Weathering:* The first step involves erosion and sedimentation of rocks and soil containing phosphates. Phosphates are very soluble, so rain and weathering can readily release phosphates from rocks, allowing the phosphate ions to find their way into the soil, and then be transported into oceans, lakes, and rivers by the movement of surface water.

- *Absorption*: Once in the soil, ocean, lake, or river, the phosphates are ingested or absorbed by animals and plants. Once inside the plant or animal, the phosphates are incorporated into its tissues and cells.

- *Terrestrial decomposition*: This step involves the remains of terrestrial plants and animals. When plants and animals die, their phosphate-rich remains are buried and compacted in the soil. In the soil, the compacted remains travel deep into the earth, where they are exposed to extreme heat and pressure. The heat and pressure form phosphate rocks deep in Earth's crust and mantle, and after millions of years, the phosphate rocks make their way back to the surface and are then ready for us to mine.

- *Marine decomposition*: This step involves the death of marine plants and animals. Their remains sink to the seafloor, become embedded and, like terrestrial plants and animals, are compacted and heated over thousands of years.

The critical factor that is missing from the graphic of the phosphorus cycle shown on the following page is a timeline showing how long it takes to produce this element. The global phosphorus cycle is measured in millions of years and is one of the slowest biogeochemical cycles.

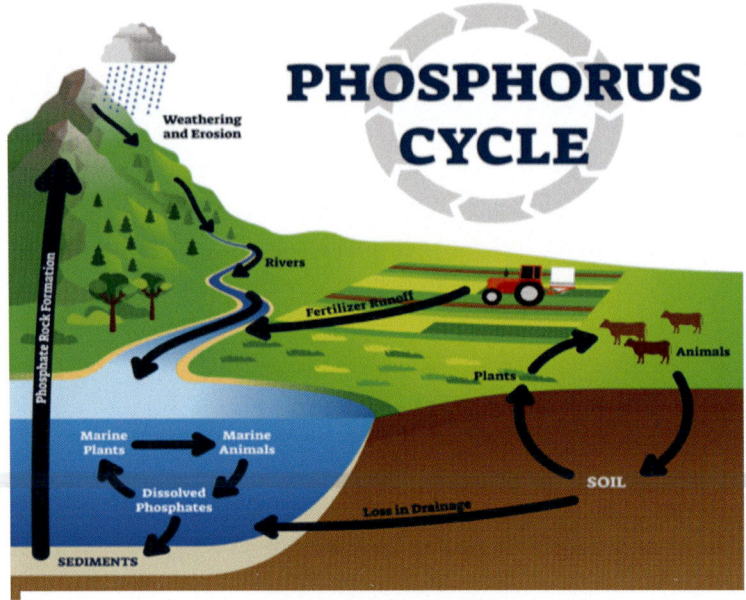

Image under license from Shutterstock

Okay, so why do we care about this? Because we make fireworks from phosphorus, and who doesn't appreciate a great fireworks show? Fireworks, however, are not the primary reason. Phosphorus-containing compounds are essential for all human and plant life. They are critical to energy metabolism, membranes, bones, and genetic compounds like DNA and RNA. As with reactive nitrogen, there is a naturally low concentration of phosphorus compounds in the soil, making phosphorus another limiting factor for food crop cultivation.

The current scale of our agriculture requires far more phosphorus than is naturally present in the soil. Therefore, farmers replenish the phosphorus pool in their soils by adding commercial phosphorus fertilizer to enrich their phosphorus-poor soil. As with reactive nitrogen, without commercial phosphorus use as fertilizer, our crop production would be about half of what it is today, and we would have already exceeded Earth's capacity to adequately feed all of us. A phosphorus compound called phosphate is mined from rock and used to make commercial fertilizer.

Phosphates dissolve readily in water, which makes it easy for them to be absorbed by plants and introduced into the food chain when eaten by humans or animals. Estimates of economically accessible reserves vary, generally a result of differing assumptions about population growth rates, global affluence, and growth of the biofuel industry. Comments from A.L. Smit, et al. of the Wageningen Institute in the Netherlands, in a detailed paper on phosphorus trends and issues caught my attention. "Now [at zero population growth] reserves would be depleted in 70-100 years and the reserve base between 170 and 264 years. Still, the risk of running out of artificial phosphorus fertilizer and the potentially fierce consequences do not feature prominently on the agenda of global UN and agricultural organizations, nor as an urgent matter on the political agenda of many countries. The governance of global phosphorus resources is left to market forces of supply and demand, and no international organizations are active in this respect."[9] It is important to keep in mind that biofuels from crops, such as corn, can usurp large amounts of phosphorus (and nitrogen) from food crops. As a result of this concern, biofuel development is moving away from food crops toward products like algae that can be fed from waste products. We will address biofuel development in Chapter 6.

Before the use of mined phosphorus fertilizer, the phosphorus cycle operated as a closed cycle, as shown in the first box on the far left in the graphic on the following page from Marissa deBoer. Crops were grown using natural fertilizer from the roots and other remains of the harvested crops, along with excrement from farm animals, who were permitted to graze on fields lying fallow or planted with grazing crops. Modern industrial level farming requires more phosphorus than these methods can provide, and we are depleting whatever phosphorus remains in the soil by clearing and tilling agricultural fields, harvest after harvest. These losses are shown by the red arrow in the middle box of the graphic. The author of this graphic envisions a new phosphorus cycle where once again, farm waste can be recycled to apply to crops, supplemented by commercial phosphorus, as shown in the box on the far right of the graphic.[10] Yep, another natural cycle we have busted. Many of the alternative solutions involve the recycling of urine and solid waste. Humans alone excrete over 3 million metric tons of phosphorus that is relatively free of harmful microbes, enough to meet about 25% of current demand. I am very excited about some of the recycled waste solutions we will review in Chapter 6.

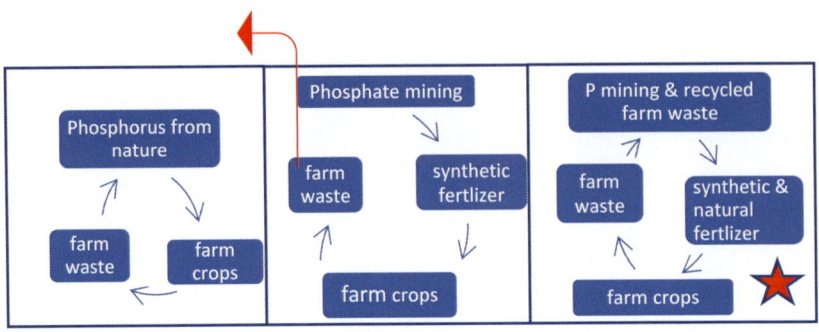

Credit: Marissa deBoer, https://www.sbs.com.au/ topics/ science/earth /article2016/ 02/19/how-great-phosphorusshortage-could-leave-us-all-hungry

There is one final issue reflected in the deBoer graphic that I need to highlight. Growing crops in the absence of synthetic fertilizer gives rise to what is referred to as the "shadow land footprint." When no synthetic fertilizer is used, yields are considerably lower, and more land is required to produce a given amount of food. Today researchers concur that about 50% of our population is being fed through the use of reactive nitrogen. If synthetic fertilizer were eliminated, we would need about twice as much cropland to produce the same yields as today. Ms. deBoer addresses this issue by proposing that our farming future include a combination of both synthetic and naturally occurring fertilizers.

Plants absorb about 30% to 50% of the phosphorus in the soil. By way of example, if a corn farmer applies 100 pounds of commercial fertilizer to the soil, 30% to 50% will be taken up and used by the corn. However, 50% to 70% of it will end up swept away by the wind or washed away, amounting to 50 to 70 pounds of phosphorus per acre ending up in our waterways. Excess phosphorus in our groundwater and surface water causes the very same reactions as excess nitrogen. Both act as nutrients that feed naturally-occurring algae, allowing the algae to multiply so fast that they commandeer most of the oxygen in the water. Deprived of oxygen, marine plant and animal life die. Some of these algae blooms produce toxins that make humans and their pets very sick when they breathe in the toxins or drink the water. In Florida, these nutrient-driven algae blooms have cost the state billions of dollars in lost tourist revenues over the years, and I have described this in great detail in the section on food crops in Chapter 5.

PHOSPHATE MINING: Okay, so we are depleting another finite resource by disrupting the phosphorus cycle. That is definitely bad. What about the method of mining phosphates to produce commercial fertilizer? Worse, much worse. I will try to condense this tale. There is an area in the center of the Sunshine State called Bone Valley. It is loaded with commercially accessible phosphates, and Florida is one of the top suppliers of phosphates in the United States (Although, Morocco and China together have 70% of the world's phosphate reserves.). Much of the phosphate-rich land in Florida is owned by one mining company. Phosphates are extracted from the ground with massive earthmovers that have a giant dragline bucket hanging by chains from a tall crane attached to the earthmover. These earthmovers are some of the largest pieces of equipment on the planet, and they look like morbidly-obese army tanks with a towering crane and an immense dragline bucket. To give you a picture of the scale of these earthmovers, the dragline bucket is large enough to hold two yellow school buses.

The operator drags the huge bucket along the ground to the first 30 or more feet of soil and sediment (you know, the vegetation and mineral-rich topsoil), and then places the topsoil aside to be used later when the site is "restored." The operator then excavates the next layer of deposits, which holds pebbles of phosphate in a matrix of clay and sand. The operator delivers each bucket of matrix to an open pit, where the matrix is mixed with water to make a slurry. The slurry is then pumped through several miles of pipeline to the beneficiation plant.

At the beneficiation plant, the slurry is sifted to separate out the phosphate pebbles. The extracted phosphate is loaded on to a rail car, where it is moved to the fertilizer production plant. The waste, which is everything that was removed from the ground except the phosphate itself, is then returned to the mining site. The earthmover operator piles the solid materials into large stacks, which can be as high as 200 feet (the height of a 20-story building) and several acres wide. The operator then delivers the slurry containing clay and other materials suspended in water to huge settling ponds. The stacks of material at the mining site are called phosphogypsum stacks. Guess what? The phosphogypsum is radioactive, at levels that would not be considered naturally occurring background levels. This level of radioactivity

31

occurs when the matrix containing naturally occurring uranium, thorium, and radium$_{226}$ is treated with sulfuric acid to make phosphoric acid for fertilizer. Like the waste from conventional nuclear power plants, phosphate companies have not resolved exactly what to do with these mountains of radioactive waste products.

The EPA requires the mining companies to remediate their mining sites, but the regulations do not require the companies to restore the land to its original pristine condition. The regulations require that the land be returned to a "usable" condition. I suspect that "usable" is intended to be a nebulous regulatory term. The best clarification I could find was from a 1975 case stating that mining companies must make the land suitable for beneficial use. There is currently a class-action lawsuit pending in Florida against a mining company that constructed a luxury housing subdivision atop a "restored" phosphate mining site.

BOTTOM LINE: PHOSPHORUS: Unlike reactive nitrogen, which we have learned to produce from atmospheric nitrogen, phosphorus cannot be produced or substituted with another substance. Our crop farming is dependent upon phosphorus to produce food at the levels needed to feed our current population. Because we have exhausted the natural phosphorus pools in our soils, farmers have resorted to using commercially produced phosphorus made from mined phosphate. As a result of a growing population and increasing demand for crop-fed livestock, we are draining our global reserves of commercially accessible phosphates. To add insult to injury, phosphate mining is proving to be extremely harmful to the mining site. That leaves us with the challenge of how we will grow enough food with the arable land we have. There are phosphate reserves under the floors of our oceans, and the idea of mining under the sea has not escaped notice. The potential environmental impact of mining phosphate rocks from the sea floor would concern any person or community economically dependent upon the sea. This is particularly true when we know that more than half of the phosphorus we are currently applying is not being used by our plants, but instead, is being washed away by surface water into our waterways. Our dependency upon a diminishing resource to grow our food is a considerable threat to the planet's biocapacity. We must learn to grow our food sustainably, with reduced use of commercial phosphates. In addition, this subject is sorely in need of serious air time to grab the attention of policymakers.

FRESH WATER AND THE WATER CYCLE

Let's start with the basics, recalling that 97% of the water on Earth is salt water, and only 3% is fresh water. Historically, more than half of this 3% fresh water has been frozen in glaciers and polar ice caps. Most of the remaining unfrozen fresh water is groundwater, with only a small fraction of fresh water located above ground in lakes, rivers, and streams or in the air as water vapor. The US Geological Survey (USGS) has estimated the supply of fresh water in lakes and rivers to be about 22,339 cubic miles and the supply of fresh water stored in the ground to be 2,500,000 cubic miles.[11] For reference, one cubic mile contains 1,101,117,147,428.57 US gallons. The recycling of water through the water cycle should keep the planet supplied with adequate fresh water through precipitation, evaporation, transpiration, and condensation. Yet, we have managed to defile this critical system, thereby reducing the biocapacity of our planet.

WATER CYCLE: Here are the steps of the water cycle in an oversimplified nutshell:

- Water from Earth's surface, including our oceans, turns from a liquid state to gaseous water vapor by evaporation and transpiration as it moves from Earth's surface and the surfaces of leaves of plants, respectively, into our atmosphere.

- While in the gaseous state, water vapor rises into the atmosphere on warm updrafts.

- However, the gaseous state is short-lived, as the cooler air at high altitudes causes the water vapor to condense into tiny particles of ice and water droplets, coming together to form clouds.

- Upper-level winds move the clouds in the atmosphere until they become saturated and release the water they hold as rain, ice, or snow.

- The process begins anew, as surface water becomes water vapor once again by evaporation or transpiration.

Natural aerosols, such as salt and dust, assist in cloud formation by providing a small seed or nucleus for the water vapor molecules to form around. Similarly, manmade aerosols, such as particulate from burning fossil fuels, are typically smaller in size and appear in much greater quantities. These seeds are called *cloud condensation nuclei.* Ironically, in large quantities, these pollution particulates help create larger and brighter clouds. Clouds that are larger and brighter block more solar radiation from striking Earth's surface, providing a cooling effect. Climate scientists suspect that this cooling effect from pollution particulates may be significant and will be seeking to quantify this effect and its impacts on regional climate patterns.

WATER CYCLE
Image under license from Shutterstock.com

In theory, the water you brushed your teeth with this morning could have contained the very same hydrogen and oxygen atoms that made up the water in a leather flask carried by Plato. The graphic above provides an excellent depiction of the steps of the water cycle.

WATER SCARCITY: The World Economic Forum (WEF) is an organization that brings together public and private stakeholders to solve the planet's most challenging economic problems. This organization considers *water scarcity* to be one of the top ten most significant global risks over the next decade.[12] Fresh water is not optional, unlike some of our other diminishing resources. Each plant and animal on the planet requires a certain minimum amount of water for biological survival. According to the World Health Organization (WHO), current fresh water scarcity has turned sanitation into a luxury commodity for about two billion people whose water sources are contaminated. Every single day, hundreds of women and children walk long distances to collect water and carry it home in containers. Approximately 800 million people currently lack essential drinking water services, including 144 million people who are dependent on surface water for drinking water. Surface water is easily contaminated and contaminated water can transmit a host of diseases, including diarrhea, cholera, polio, and typhoid.

The United Nations Water agency (UN-Water) estimates that 50% to 64% of the world's population currently lives in water-stressed regions. The water consumption disparity between the United States and other countries, both developed and less-developed, is stark: According to the EPA, the average family in the United States uses about 300 gallons of water a day at home. Although, other sources have calculated actual consumption to be closer to 1,500 gallons per day after you factor in meat and dairy products eaten and manufactured products owned. The graphic on the following page shows the amount of renewable fresh water, by country in cubic meters. As you can see, a large swath of our global population has limited water supplies available to them, particularly in areas of the African and South Asian continents. Water stress is defined by UN-Water as less than 1,215 US gallons of fresh water per day per person, water scarcity as less than 713 US gallons per day per person, and absolute water scarcity is defined as less than 370 US gallons per day per person. By these measures, 49 countries are presently water stressed, 21 of which are suffering absolute water scarcity.[13]

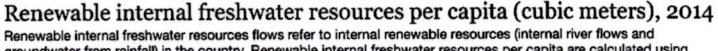

Renewable internal freshwater resources per capita (cubic meters), 2014
Renewable internal freshwater resources flows refer to internal renewable resources (internal river flows and groundwater from rainfall) in the country. Renewable internal freshwater resources per capita are calculated using the World Bank's population estimates.

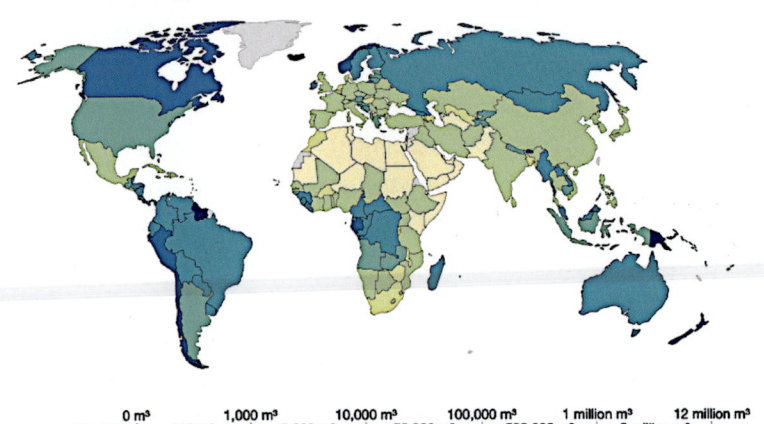

Source: World Bank

OurWorldInData.org/water-access-resources-sanitation/ • CC BY

Image licensed by Our World in Data, Creative Commons License by Attribution

It continues to astonish me that much of the developed world uses potable drinking water to flush waste from toilets, wash cars, and water lawns when we could be using gray water. Why aren't new houses built with gray-water systems, so washing machine and dishwasher water can be used at least for irrigation or toilet flushing? Why are we putting precious drinking water on the azaleas? There was another version of the graphic above that depicts the cubic meters of fresh water supplies per capita over time for each country, each region, and the world. The chart shows a continued decrease in the amount of globally available fresh water, and also reflects that the amount of available freshwater per capita in 2014 was one half of the supply of freshwater per capita in 1962. The hydrogen and oxygen atoms making up our water are supposed to keep recycling indefinitely. Water is not supposed to be a diminishing resource!

Fresh water is a critical driver of industry, right up there with fossil fuels and other energy sources, and it is one of the key ingredients for agriculture, without which we cannot eat. In the United States, 5% of all water consumed is by homes and businesses, 55% is consumed by agriculture, and the remaining 40% is consumed by industrial applications, including power generation. So, even if we have enough water for drinking and sanitation, without adequate fresh water supplies, the wheels of industry and food production quickly grind to a halt.

GROUNDWATER DEPLETION: Groundwater is a valuable resource throughout the world. Groundwater comes from aquifers, which are gravel- and sand-filled underground freshwater reservoirs. We only see this water when we pump it out of the ground, or when it flows to the surface in a natural spring. The most accessible water is surface water, such as lakes and rivers, but as surface water sources become increasingly scarce, we are accessing our groundwater supplies more and more. Groundwater is the source of drinking water for nearly all of the rural population in the United States and about 50% of the non-rural population.[14] Groundwater is the most extracted natural resource, most of which is directed to agriculture. At the same time, demand for fresh water is increasing around the globe, driven by global population growth, the expansion of irrigated agriculture, and economic development. This increasing demand is largely being met by groundwater.

Under typical conditions, aquifers recharge when rain and surface water from streams seep into underground reservoirs. Groundwater depletion occurs when we pump water out of the ground faster than it is recharged, causing the volume of groundwater in storage to decrease. Groundwater depletion can have widespread effects, including the following: (i) it can cause wells in the immediate area to dry up when the water table is drawn down below the depth of the wells, (ii) it can reduce surface water reserves of lakes fed by groundwater, and (iii) it can cause water quality to diminish if saltwater contaminates the fresh groundwater. In some areas, deep below the fresh groundwater is saltwater. With normal levels of pumping, the boundary between the fresh water and salt water tends to be relatively stable; however, excessive pumping can cause saltwater to migrate inland and upward and cause saltwater contamination of the freshwater supply.

According to a group of scientists from the American Geophysical Union, a nonprofit organization established in 1919 to promote scientific research for the benefit of humanity, the rate at which we are pumping the water stored in the ground has doubled since 1960. We are globally depleting our groundwater faster than normal rainfall can recharge aquifers.[15]

The map below from the United States Geological Survey (USGS) shows areas of groundwater depletion in the United States, much of which is located in the arid Southwest and the agricultural Midwest. Data from one of NASA's GRACE satellites suggest that 13 of the world's 37 most significant aquifers are being seriously depleted by irrigation and other uses at a much faster rate than they can be recharged by precipitation or runoff. Consequently, even though the water cycle should be recycling water molecules so that we have a stable supply of water, fresh groundwater is gradually becoming a nonrenewable resource in areas where aquifers are unable to recharge because of overuse.

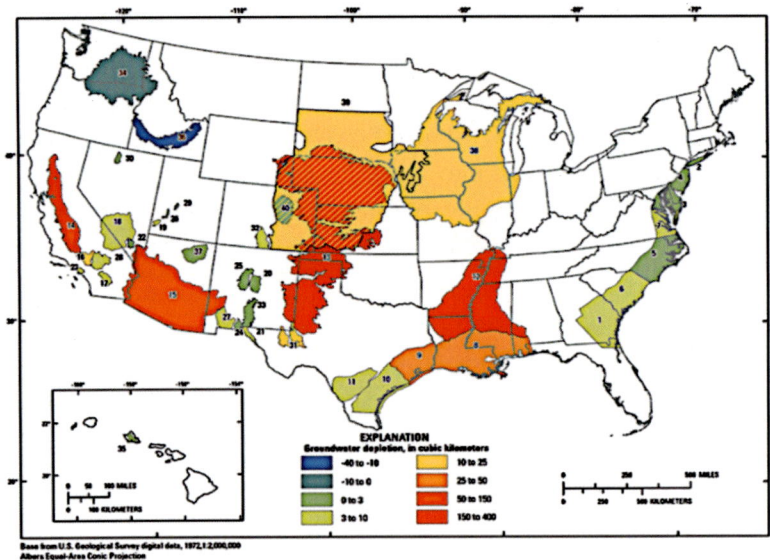

Figure 2. Map of the United States (excluding Alaska) showing cumulative groundwater depletion, 1900 through 2008, in 40 assessed aquifer systems or subareas. Index numbers are defined in table 1. Colors are hatched in the Dakota aquifer (area 39) where the aquifer overlaps with other aquifers having different values of depletion.

The table below is a list of some consumer goods, showing their respective water footprints. I dare you to measure the water footprint of your clothing and other personal items. I did just that, and the result was astonishing. Go to the Water Footprint Network website cited at the bottom of the table to calculate the water footprint of other consumer products.

Water Footprint of Common Consumer Items

Item	Water Footprint
Car	13,737 – 21,926 gallons
Leather Shoes	3,626 gallons
Smart phone (mobile)	3,190 gallons
Bed Sheet (cotton)	2,839 gallons
Jeans (cotton)	2,108 gallons
T-shirt (cotton)	659 gallons

Sources: Berger et al., Friends of the Earth/Trucost, Water Footprint Network
https://www.watercalculator.org/footprint/the-hidden-water-in-everyday-products

TRANSBOUNDARY WATERS: Water knows no national boundaries, which can raise significant geopolitical complications in trying to achieve universal water access. UN-Water is an agency of the United Nations tasked with ensuring universal access to water and sanitation services. This agency has reported that there are 263 transboundary lake and river basins covering over 46% of Earth's surface and approximately 300 transboundary aquifers. It is easy to see how these resources can lead to conflicts over water, particularly in water-deficient areas. As a hypothetical example, does country X, which is upstream from country Y, have the right to dam a river that flows through country X into country Y? According to UN-Water, there have been 37 serious conflicts over water in the past 70 years, but 295 international water agreements have been executed during the same period. When you contemplate the plight of small countries that are downstream of a larger, more affluent country, the potential for conflict is not difficult to imagine. According to the UN-Water, about

66% of the transboundary rivers do not have a cooperative management agreement in place.[16]

WATER POLLUTION: No discussion of water would be complete without at least a nod to water pollution, one of the early environmental issues to come into the collective consciousness back in the 1970s. Pollution of our water sources reduces the amount of water available to fuel the activities of our global population, thereby reducing the planet's biocapacity. I am going to provide an abbreviated and restated inventory of water pollution issues for your consideration, from the National Resource Defense Council (NRD). The NRDC is a nonprofit environmental advocacy organization active since 1970. [17]

- *Agriculture:* Agriculture is a water hog, consuming about 70% of Earth's freshwater supplies, and it is also a significant source of water pollution. Food crop agriculture and livestock agriculture are the primary sources of degradation of our groundwater and surface water supplies. When it rains, fertilizers, metals, pesticides, pharmaceuticals, and animal waste from farms and livestock operations wash chemicals and pathogens into our waterways. Livestock operations on land have been implicated in creating more than 500 dead zones in our oceans. In the Gulf of Mexico, just south of Louisiana, lies a 7,000-square mile dead zone created by the drainage from the Mississippi River watershed. In addition, soil and sediment that is washed off agriculture fields are injurious to nearby lakes and streams when they cloud the water enough to reduce the amount of essential solar energy that can reach aquatic plants. Finally, seepage from manure management systems can reach ground water aquifers, bringing viruses and bacteria to water supplies.

- *Oil and Gas*: While major oil spills, like BP's Deepwater Horizon spill, definitely grab our attention, according to a 2002 report by the US National Research Council of the National Academy of Sciences (NRC-NAS), most of the oil pollution in our seas is not from these accidents. According to the NRC-NAS, of the 29 million gallons of petroleum that enter North American oceans each year due to human activities, less than

10% comes from tanker and oil pipeline spills, and about 3% comes from exploration and extraction activities. Most of the rest comes from land-based runoff from farms, urban areas, industry, and coastal refineries. Interestingly, natural oil seeps at the seafloor are reported by the NRC-NAS as the source of nearly half of all oil pollution in North American seas. I am a bit skeptical about the actual percentages attributed to the various oil pollution sources in this report, but I found other reports at variance with this report and at variance with each other. Whatever the share attributed to each pollution source, scientists seem to agree that about 1.3 to 2.3 *million* tons of oil end up in our global oceans each year.

- *Wastewater*: Used water is wastewater. It comes from our sinks, showers, and toilets (think sewage) and from commercial, industrial, and agricultural activities (think metals, solvents, and toxic sludge). The term also includes stormwater runoff, which occurs when rainfall carries road salts, oil, grease, fertilizers, chemicals, and debris from impermeable surfaces into our waterways. More than 80% of the world's wastewater flows back into the environment without being treated, according to UN-Water; in some lesser-developed countries, the figure exceeds 95%.[18]

- *Radioactive Waste:* Radioactive waste is any pollution that emits radiation beyond that which is naturally released by the environment, which is called the "background level." It is generated by uranium mining, nuclear power plants, military weapons testing, and universities and hospitals that use radioactive materials for research and medicine. Radioactive waste can persist in the environment for thousands of years, making disposal a major challenge.

BOTTOM LINE: WATER CYCLE: Fresh water should be an infinitely renewable resource. There should exist the same amount of fresh water on earth today as existed when the sharpest pencil in the box was bacterial slime. As our population grows, pressure on our limited supply of fresh water is increasing, resulting in less fresh water being available for drinking, sanitation, agriculture, and production of goods. Additionally, we are polluting both surface water and groundwater, and we are doing so at a rate that does not allow it to cleanse itself through natural systems. In many locations, we are consuming more fresh water than Earth's natural systems can replenish.

ARABLE LAND AND SOIL

A SCARCE RESOURCE: Arable land is land that is suitable for the cultivation of crops. Let's begin with the fact that 71% of the planet is ocean, leaving humankind and land-based plants and animals with the remaining 29% upon which to make a life. But, wait. Before you start planting, we need to remove 8.7% of the planet that is covered by glaciers and ice sheets or is barren, plus 10.2% of the planet that is forested, plus 0.3% of the planet that is surface fresh water, and finally, 0.3% of the planet that is urban. That leaves us with 9.5% of the planet as potentially arable land available for cultivation of crops.

Today, we use 77% of that arable land to raise crops for livestock and to graze livestock, leaving us with 23% of that 9.5% to grow food crops for humans (about 2.19% of the planet). That minuscule portion of arable land for food crops can barely provide sufficient calories for our current population, let alone the increased population we expect to have by midcentury. Along with a paucity of arable land, we are losing approximately 75 million acres of rich topsoil from arable land each year due to soil erosion.[19] Please see the graphic on the following page for a clear picture of how little land we actually have to grow our food. We should also consider that this tiny portion of the planet that is available for food crops is not evenly distributed across the globe. Consequently, in areas of Amazonian South America and sub-Saharan African countries, lush forests are being slashed and burned to create additional cropland to grow livestock crops and for livestock grazing. Additionally, not all of the planet's arable land is equally near water for irrigation, nor is it equally fertile.

As stated earlier, research has estimated that without commercial fertilizers, we could only feed about 50% of the world's population. Additionally, while global population more than doubled between 1960 and 2010 (3 billion to 6.8 billion), the amount of arable land increased only 4% during that same time. The primary reason arable land increased at all during that period is because we have started to slash and burn forests to create agricultural land. This is a classic Malthusian dilemma: geometric population increase with arithmetic land increase.

Global surface area allocation for food production

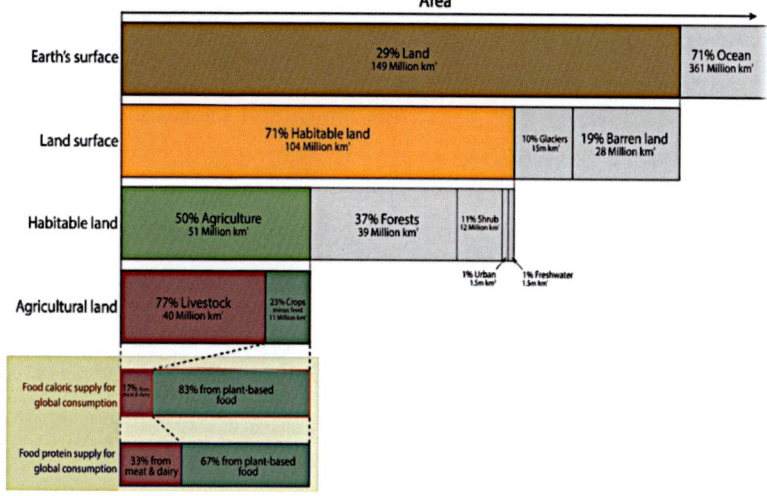

Image licensed by Our World in Data, Creative Commons License by Attribution

TOPSOIL: We cannot leave the subject of arable land without addressing topsoil. Soil is one of the most critical life-sustaining natural resources. It provides the nutrients for food production, performs carbon sequestration, provides habitat for plants and animals, and regulates the absorption of surface water. Yet, in writing this, I realized how easy it is to take topsoil for granted, as an infinite resource that will always be there for us. Even our slang expressions show how little regard we have for our soil: "dirt-poor," "dirt-cheap," "treated like dirt," or "to dig up dirt" all have negative connotations. Well, what is soil, and how did it get in my backyard?

Soil is formed from rock disintegrating into sediment as a result of wind, water, chemical reactions, and temperature changes. Sediment from disintegrating rock is transported and accumulates with the help of wind, water, and gravity. The original minerals in soil are from the parent rocks from which the sediment came and will determine many of the properties of the soil. Accumulated mineral-rich sediment invites plants, burrowing insects, and microorganisms such as bacteria and fungi to take up residence. Over time, these plants and other organisms mature and die, and new ones replace them, with each cycle adding leaves, roots, and other nutrient-rich organic debris to the soil. Animals may graze on the plants, and their waste is broken down and added to the sediment. Bacteria, fungi, and worms continue to break down the animal waste and plant and animal remains, adding more nutrients to the sediment, until voilà, a few hundred years later, you have an inch of topsoil! (This timeline is more like thousands of years if you factor in the rock cycle and the time it takes to form the rocks from which the soil originates.)

A cup of soil contains millions of varied species of bacteria and fungi, microscopic single-celled protozoa, and nematodes (worm-like insects). We are eroding our topsoil as much as 40 times faster than it can naturally form.[20] Historically farmers employed methods that nurtured the soil, knowing that a robust soil ecosystem meant they could produce sufficient food. Today we are depleting our soils of nutrients and using methods that lead to increased erosion. By way of example, plowing turns over the soil, mixes it with air, and stimulates decomposition of its organic matter. Over time, this leads to a reduction in soil nutrients and the rich organic debris in the soil and reduces the ability of the soil to absorb atmospheric carbon dioxide, retain water, and resist erosion. As early as 1943, the idea of minimum tillage was introduced, but it did not take hold until the 1970s, a period of active conservation awareness. By 1997, almost 37% of our cropland was being managed with minimal tilling. Between 1982 and 1997, minimal tillage produced a 42% drop in erosion and a significant increase in organic material in the soils. Since 1997, it appears that minimum tillage practices have not increased. The move to industrial farming has increased the need for tilling, as the topsoil and subsoil become compacted from the use of four- to ten-ton harvesters and other equipment.

BOTTOM LINE: ARABLE LAND: We have more and more people needing to be fed with less and less per capita fertile and irrigated arable land. This is sharply diminishing the capacity of the planet to adequately feed the current population. Remember also that soil is one of our excellent carbon sinks that removes atmospheric carbon dioxide, but aggressive tilling sends all that stored carbon back into the atmosphere and reduces the soil's ability to sequester carbon. If we expect to feed the additional several billion humans expected to join us over the rest of this century, we must begin to once again view our arable land as an independent ecosystem and begin to universally employ agricultural methods that allow the components of this ecosystem to flourish and regenerate. The term for the research and development of these best agricultural practices is "agroecology," and the term used for the implementation of these best practices is "regenerative agriculture."

RARE EARTH ELEMENTS

Rare earth elements are not particularly rare, but they possess unique properties that make them essential to many modern optical, magnetic, medical, electronic, and catalytic technologies. They consist of a group of 17 elements that occur together toward the end of the periodic table, including yttrium, lanthanum, cerium, praseodymium, neodymium, promethium, samarium, europium, gadolinium, terbium, dysprosium, holmium, erbium, thulium, ytterbium, scandium, and lutetium. The rare earth elements are metals that share many characteristics, and they are often found together in geologic deposits. I can't pronounce most of these, but I hope we will continue to extract these minerals, since they are used in computer memory chips, DVDs, rechargeable batteries (like those in your electronic devices and your hybrid or electric car), smartphones, catalytic converters, magnets, medical imaging devices, and fluorescent lighting, to name a few. They are even used in national defense as an essential component of night vision goggles, guidance systems, and precision-guided weapons. The chart on the following page lists all of the rare earth elements, along with common usages and production sources.

According to the USGS, these metals are in relatively abundant supply throughout the world; however, they are very challenging (read: expensive) to mine because it is unusual to find them in concentrations high enough to justify the cost of extraction. As you will see in the graphic, China controls the most substantial reserves of rare earth elements, so we should try to play nice.

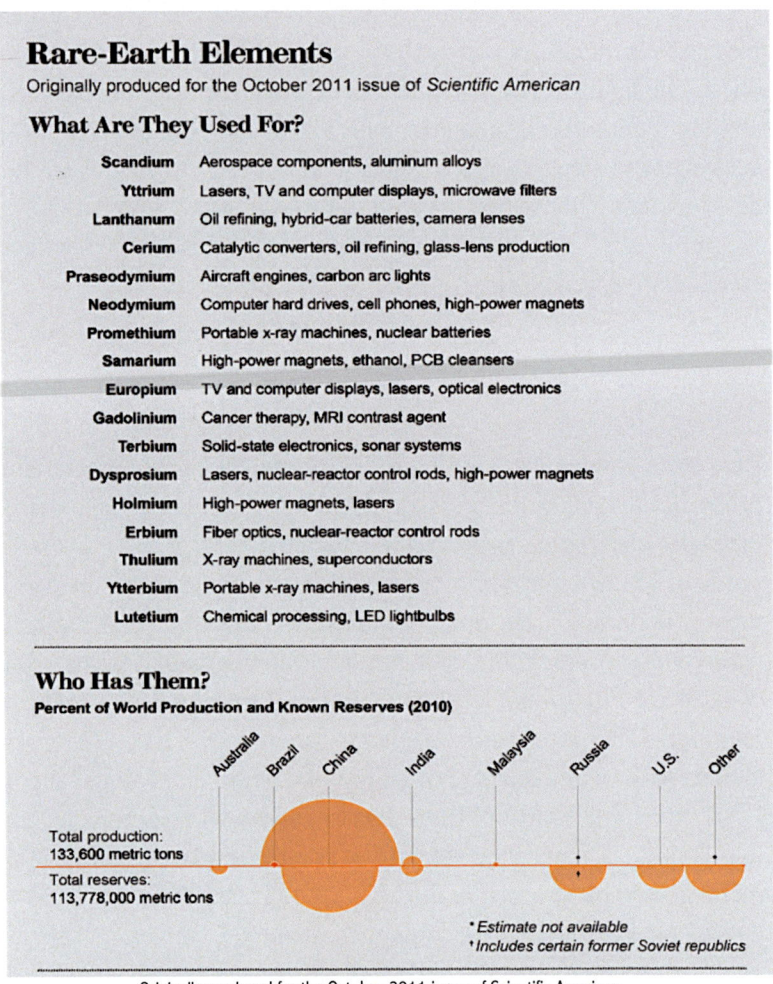

Rare-Earth Elements

Originally produced for the October 2011 issue of *Scientific American*

What Are They Used For?

Scandium	Aerospace components, aluminum alloys
Yttrium	Lasers, TV and computer displays, microwave filters
Lanthanum	Oil refining, hybrid-car batteries, camera lenses
Cerium	Catalytic converters, oil refining, glass-lens production
Praseodymium	Aircraft engines, carbon arc lights
Neodymium	Computer hard drives, cell phones, high-power magnets
Promethium	Portable x-ray machines, nuclear batteries
Samarium	High-power magnets, ethanol, PCB cleansers
Europium	TV and computer displays, lasers, optical electronics
Gadolinium	Cancer therapy, MRI contrast agent
Terbium	Solid-state electronics, sonar systems
Dysprosium	Lasers, nuclear-reactor control rods, high-power magnets
Holmium	High-power magnets, lasers
Erbium	Fiber optics, nuclear-reactor control rods
Thulium	X-ray machines, superconductors
Ytterbium	Portable x-ray machines, lasers
Lutetium	Chemical processing, LED lightbulbs

Who Has Them?

Percent of World Production and Known Reserves (2010)

Australia Brazil China India Malaysia Russia U.S. Other

Total production: 133,600 metric tons

Total reserves: 113,778,000 metric tons

* Estimate not available
† Includes certain former Soviet republics

Originally produced for the October 2011 issue of Scientific American.
Credit: Jen Christiansen Source: Mineral Commodity Summaries 2011, USGS
https://www.scientificamerican.com/article/dont-panic-about-rare-earth-elements/

Once extracted, rare earth elements require significant processing to become usable, as described in a *Scientific American article in 2019*.

> [They require] complex chemical processes that typically involve a procedure called solvent extraction, in which the dissolved materials go through hundreds of liquid-containing chambers that separate individual elements or compounds— steps that may need to be repeated hundreds of times. Once

purified, the material can be processed into oxides, phosphors, metals, alloys, and magnets that take advantage of these elements' unique magnetic, luminescent, or electrochemical properties. The strong and lightweight nature of rare earth magnets, metals, and alloys have made them especially valuable in high-tech products.[21]

While the United States holds only modest reserves of rare earth elements, in far West Texas, there is a 1,250-foot mountain called Round Top, located south of El Paso, TX. Texas Mineral Resources Corp. has a development agreement with USA Rare Earth, to extract and process the rare earth elements and other technology metals and industrial minerals from this site. In addition, the project is expected to include a pilot processing facility in Colorado, to separate and purify the extracted minerals. According to the company's recent announcement, it expects to produce 1900 tons of rare earth minerals per year for about 20 years, extracting only 14% of the ore. The mountain is located on State property, and I am hoping that the 6.25% royalties that will be paid to the Texas General Land Office from the project will not get in the way of strict enforcement of environmental requirements.

Environmental scientists have expressed concerns that rare earth mineral mining and production can result in tailings ponds that contain acids, heavy metals, radioactive elements, and other toxins that can leak into the groundwater. The enormous demand for these elements could create a strong temptation to compromise environmental requirements during extraction. China is selling its rare earth minerals at very competitive prices that reflect its lax environmental requirements. One of China's largest mining sites, Bayan-Obo, has created a huge tailings pond that is 11 square kilometers, about the size of three New York Central Parks. The toxic sludge from this tailings pond is producing elevated concentrations of radioactivity. The environmental mess that mining for rare earth minerals creates is ironic in an unfortunate way, as many of the technologies using these elements are essential to our new green economy (e.g., wind turbines, electric cars, rechargeable batteries). According to the United Nations International Resource Panel (IRP), only *1%* of rare earth elements are being recycled.

BOTTOM LINE: RARE EARTH ELEMENTS: Rare earth elements are critical for certain technologies and common consumer products upon which we rely, such as smartphones, computer screens, and rechargeable batteries. The cost of extracting and processing these elements makes it almost cost-prohibitive to bring them to market. Compounding the cost of extraction and processing is the fact that current extraction methods are leaving behind toxic tailings that must be cleaned up to make the extraction of these elements sustainable. In theory, increased recycling could provide a steady stream of product. However, one significant obstacle to recycling is the cost of the chemical processing required to purify the material for reuse. Researchers at the University of Pennsylvania are working on a new process to make this step less expensive, hopefully opening the door for increased recycling of rare earth minerals. Meanwhile, remember not to toss your spent electronics on to the landfill.

NONRENEWABLE METALS

While rare earth metals are in good supply, there are other critical metals that either have a finite supply or take eons of time to form. Many of our metals were formed 13.5 billion years ago during the formation of our universe in the very hot environments of our earliest stars. We discussed these early stars in the definition of biogeochemical cycles in Chapter 1. Most of the metals we extract are found in Earth's lithosphere, frequently embedded in rocks. To extract the metals, the rocks or ores must be mined and then refined or processed. If humanity were to exhaust any of these crucial and useful minerals, it would be nearly impossible to replace them.

- *Aluminum:* Aluminum is the most prevalent metal on Earth and makes up 8% of Earth's crust. Because it is lightweight, durable, reflective, easily formed, and corrosion-resistant, aluminum is used to manufacture many common items. Its versatility makes it nearly a miracle metal. Pure aluminum does not naturally occur but is found in nature as various aluminum-containing compounds in rock. After the rock is minded, it must undergo several processes before becoming serviceable aluminum we can use. According to the US Aluminum Association, about 50% of all aluminum is being recycled. When we recycle aluminum, the energy costs to make aluminum stock are only 5% to 10% of the energy cost of mining and producing the original material. Aluminum is 100% recyclable, making this metal one of the most recyclable of all materials. Recycled aluminum retains its properties indefinitely, and it is one of the materials in the waste disposal stream that more than pays for the costs associated with collecting and sorting. A large percentage, perhaps as much as 70% of all aluminum mined and processed is still in circulation.

- *Copper:* Copper is used in a variety of construction, power, and electronic products. It is highly conductive of heat and electricity and is the third most-consumed metal in industry today. Only aluminum and iron have higher consumption rates. Around 200 years ago, copper nuggets could be found in nature in certain parts of the world, but since we have extracted most of the easily accessible copper, we now mine copper from ore. This means that we strip mine areas holding a supply of rocks with embedded copper compounds, and then chemically process the rocks to separate the copper. As you might guess, this kind of mining and processing requires more energy and is more costly than bending down to pick up a copper nugget. As with aluminum, copper stock made from recycled copper is less expensive and requires only 10%-15% of the energy required to make freshly mined copper stock. Again like aluminum, copper by itself or in any of its alloy states is entirely recyclable and can be processed over and over again with no loss of quality (except for wiring, which requires mined copper that has not been previously used). Like aluminum, much of the mined copper is still in circulation. The recycling rate of copper is only about 34% in the United States.

- *Iron and Steel:* Iron makes up about 5% of Earth's crust and most of Earth's inner and outer cores as an iron-nickel alloy. Iron is found in the sun and stars of our solar system, and most iron is extracted from hematite and magnetite. Most of the iron mined today is processed into steel, which is used extensively in industry, civil engineering, and commercial

49

construction. Like copper and aluminum, steel can be recycled an infinite number of times. Making steel stock out of recycled material saves 75% of the energy required to make steel from freshly mined materials. According to the RSC, as much as 40% of worldwide steel production uses recycled iron. Steel is one of the most recycled materials, with rates as high as 78%, and it does not lose its strength or other characteristics with multiple recycling.

- *Silver:* Silver is a precious metal used in making jewelry, coins, photography, and tableware. It also has industrial use in making mirrors, printed circuits, dental alloys, solder, brazing alloys, electrical contacts, and batteries. It is highly conductive of heat and electricity and very reflective. Each year, 22,000 tons of it are produced, but only about 30% is recycled. According to research from 2013, the estimated recoverable reserves of silver are 2.7 to 3.1 million tons, of which 1.35 to 1.46 million tons have already been mined. The researchers concluded that peak silver production will occur before the middle of this century, and by 2240, all silver mines will be depleted.[22] Your spent electronics are a great source of silver.

- *Gold*: According to the RSC, 78% of mined gold is used to make jewelry. The rest is used to make bullion, coins, electronic components, computer components, dentistry accessories, and aerospace components. It is a good conductor of electricity and heat, is malleable, and corrosion-resistant. It is also a good reflector of infrared radiation and can be used to shield spacecraft from the sun's heat. There are few minerals that provide the same chemical and physical performance and reliability as gold, particularly in electronics. We have been mining and refining gold for thousands of years, and the World Gold Council estimates that only 30% of our total reserves remain. The recycling rate of gold is about 1%. Again, don't throw your spent e-waste into the landfill-- there may be gold lurking within.

- *Uranium*: Whatever your thoughts on the safety of nuclear energy, most would agree that nuclear energy does not produce greenhouse gases. It is "clean energy," and each kilogram of uranium$_{235}$ engaged in a fission reaction releases as much energy as 2,700 tons of coal. Uranium$_{235}$ is a nonrenewable resource, and it is obtained by mining. Mining activities are driven by nuclear-plant demand for fuel for the 98 reactors currently in operation in the United States and the 450 reactors worldwide (and for those under construction or planned for use in the future). In 2015, reserves were estimated at 5.7 million tons, which amounts to 135 years of supply at 2014 rates of consumption. Including less accessible reserves, would bring the estimate or reserves to 7.6 million tons—much more on using spent uranium and safer nuclear energy from thorium in Chapter 6.

BOTTOM LINE: METALS: The world's metals are not renewable resources, yet many find their way on a linear path to the landfill. Many metals can be recycled almost indefinitely without changing their chemical properties. The cost savings of using a recycled metal over a newly processed metal can be as high as 95%. Some of the most highly recycled metals are the ones discussed here, leaving too many more which end up in the landfill.

BIODIVERSITY LOSS

Extinction of plant or animal species can be local or global and is caused naturally or by human activity. Anthropogenic species loss is caused by destruction and fragmentation of habitat resulting from land development, deforestation, urbanization, the introduction of invasive species, climate change (heat stress and drought), and pollution. Scientists refer to naturally occurring species extinctions as the "background extinction level" and calculate the background extinction level to be one to five species per year. Today we are experiencing species loss at a rate of 1,000 times the background extinction level. Researchers have concluded that 99% of the species currently facing extinction are at risk as a result of human activity. To compound the problem, when a loss occurs low on the food chain, such as in a coral reef system, a trophic extinction occurs when each of the animals in the food web loses its food source. We are ravaging wildlife populations. According to the World Wildlife Fund, the number of mammals, birds, reptiles, amphibians, and fish on our planet is less than half what it was 50 years ago. [23]

Why does this matter? Biological diversity in plants and animals ties all living things, including us, into an interdependent ecosystem. The land, water, atmosphere, minerals, plants, and animals together form the web of life upon which we all rely. As we discussed in Chapter 1, when more plant and animal species exist and are thriving, existing ecosystems have a much better chance of sustaining life on the planet. Biological diversity involves genetic diversity within a given species, between species, and between communities and ecosystems.

In a March 2018 article in *The Guardian* by Damian Carrington, he described biodiversity in a way that resonated with me, although the quote may not be properly attributed to Mr. Carrington. "A more philosophical way of viewing biodiversity is this: it represents the knowledge learned by evolving species over millions of years about how to survive through the vastly varying environmental conditions Earth has experienced. Seen like that, experts warn, humanity is currently 'burning the library of life.'"[24] Loss of biological diversity is also an economic matter, affecting individual jobs, entire industries, available medicines, and food security. The World Economic Forum has placed biodiversity loss as another one of the top ten global risks facing our world.

Around 10,000 years ago, *Homo sapiens* composed 1% of the entire animal biomass of the planet. The remaining 99% was composed of a wide variety of mammals, insects, birds, crustaceans, amphibians, and fish. Animals ran wild in our pristine waters, forests, mountains, and plains. Today, 10,000 years later, a short time in the long history of our planet, humans and the livestock owned by humans make up 96% of Earth's animal biomass, and Earth's wild animals comprise the remaining 4%. Humans and their organic accoutrement have gone from making up 1% of Earth's biomass to 96%—the implications are staggering.

Chapter 3

GREENHOUSE EFFECTS

Although this is not intended to be a book about climate change, we cannot complete our discussion of the broader sustainability issues without taking a hard look at the remaining essential biogeochemical cycles of our ecosystem. The remaining cycles we will discuss are those relating to gases implicated in causing greenhouse effects. Most of the public discourse pertaining to the environment has been directed to excessive fossil fuel emissions, while certain other equally sinister gases have not been given their fair share of air time. We will put all these gases on full display in this chapter, including carbon dioxide, methane, nitrous oxide, and water vapor.

CARBON AND THE CARBON/OXYGEN CYCLE

Let's begin with a deep dive into the carbon cycle and the impact of our disruptions of this cycle. The carbon cycle is one of our most valuable natural biogeochemical cycles, and carbon is one of the most critical elements sustaining life. You and I are nothing if not walking carbon depositories, with carbon being present in every one of our cells, and in all lipids, proteins, carbohydrates, and nucleic acids. The steps depicted in the graphic of the terrestrial carbon cycle on the following page are described below, along with the steps of the parallel marine carbon cycle. I have included the marine carbon cycle because the marine cycle is as important as the terrestrial cycle in storing carbon and emitting oxygen. The terrestrial steps are shown in green, and the marine steps, somewhat abridged, are shown in blue.[25]

- Carbon from anthropogenic and natural sources moves from Earth's surface into the atmosphere, bond with oxygen, and become carbon dioxide molecules.
  Carbon from anthropogenic and natural sources moves from the oceans into the atmosphere as carbon dioxide molecules.

- Carbon in carbon dioxide molecules moves from the atmosphere to terrestrial plants when the plants absorb atmospheric carbon dioxide molecules as part of their photosynthesis processes.
  Similarly, carbon in carbon dioxide molecules moves from the atmosphere to the ocean when seawater absorbs atmospheric carbon dioxide. Marine plants in the upper portions of the sea, such as microscopic phytoplankton, absorb dissolved carbon dioxide as part of their photosynthesis process.

- Carbon moves from terrestrial plants to terrestrial animals when animals eat the plants.
  Carbon moves from phytoplankton to zooplankton (microscopic marine animal life) and to larger marine organisms when marine animal organisms eat the marine plants.

- Carbon moves from terrestrial plants and animals to the soil when plants and animals die, and carbon compounds in their bodies and structures enter the soil.
  Carbon moves from the upper ocean to the sea floor when marine plants and animals die and sink to the sea floor.

- Decomposers like bacteria and fungi in the soil break down the carbon compounds from the dead plants and animals.

Carbon also moves from the deep ocean to the upper ocean as dissolved carbon dioxide when decomposers break down carbon compounds from dead marine plants and animals.

- Terrestrial plants release oxygen into the atmosphere as part of the plants' respiration and photosynthesis.
Marine plants release oxygen into the sea as part of their respiration and photosynthesis.

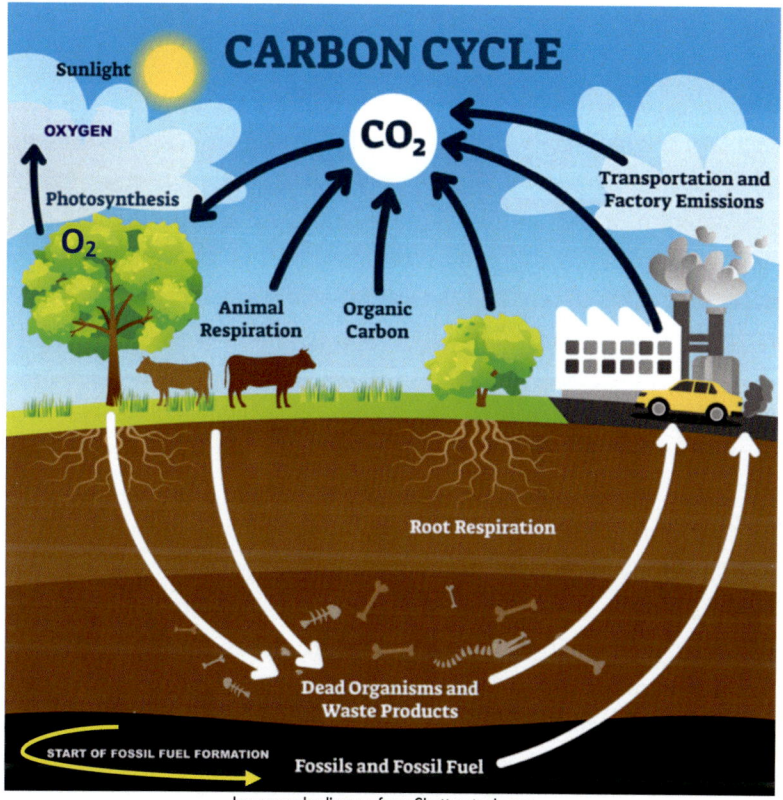

Image under license from Shutterstock.com

In discussions of carbon and carbon dioxide, you will usually encounter two terms: "carbon sink" and "carbon sequestration." A *carbon sink* is a reservoir, either natural or man-made, that can absorb and store carbon compounds for an indefinite period, such as our soil, oceans, and plants. A carbon reservoir is considered a carbon sink if it absorbs more carbon than it emits. *Carbon sequestration* is the process of capturing and storing carbon in a carbon sink for an indefinite period. For instance, forests sequester carbon by capturing carbon dioxide from the atmosphere and transforming it into biomass through photosynthesis. Prior to industrialization, the carbon budget was "balanced," as the amount of carbon emitted into the atmosphere by natural and anthropogenic sources was equal to the amount of atmospheric carbon our sinks were capable of sequestering. The build-up of carbon dioxide currently in the atmosphere is a result of the disruption of this cycle. The existing carbon sinks cannot absorb all of the carbon we have emitted and continue to emit. The result is a net increase in atmospheric carbon dioxide. Take another look at the carbon cycle graphic to identify the various carbon sources and sinks.

Joint research from the University of California and the American Meteorological Society has suggested that projections of greenhouse gas concentrations may be materially understated because carbon sequestration in soil and terrestrial plant life may not increase at the same rate as carbon dioxide concentrations in the atmosphere are increasing; therefore, our terrestrial carbon sinks may not be functioning at levels assumed by recent greenhouse gas projections.[26] Research published in early 2020 supports this conclusion, particularly with regard to the Amazon forests, and attributes the reduction in sink capacity to the degrading health of the trees due to climate change. The research concluded that our forest sink capacity peaked in the 1990s.[27] Both research findings suggest that the carbon cycle may have limits that we have not fully appreciated. Could there be a "saturation point" for our carbon sinks, beyond which they will not sequester any more carbon? If our carbon sinks are failing to draw in carbon dioxide at assumed historical rates, then atmospheric carbon dioxide will accumulate faster than estimated, and our mitigation efforts will be less effective than projected.

When we burn coal or eliminate forests by burning them, substantial amounts of carbon that had been stored in the coal and the trees are released into the atmosphere. The released carbon atoms bond with oxygen atoms to become molecular carbon dioxide, taking up residence in our atmosphere. When there is more carbon dioxide in the atmosphere than our oceans, plants, trees, marine plant life, and soil (e.g., carbon sinks) can absorb, the excess carbon dioxide becomes trapped in Earth's atmosphere for a century or longer, and the result is a severely disrupted carbon cycle. A healthy carbon cycle would have the ability to recycle all of the carbon released on this planet over and over, indefinitely. This infinite recycling cannot be effective unless there is a balance between the carbon dioxide being emitted into the atmosphere and the carbon dioxide that can be absorbed by plant life, soils, and our oceans.

It is important to remember that not all of the carbon dioxide in the atmosphere and oceans is a result of human activity. Some of our atmospheric carbon dioxide is emitted naturally from various sources, such as dead terrestrial plants and animals, respiration of animals, ocean emissions, volcanic eruptions, decomposers in the soil like bacteria and fungi, and natural carbon deposits within Earth's crust. For millions of years, the amount of carbon dioxide produced by these natural sources has been offset by Earth's natural carbon sinks. For the hundreds of thousands of years before industrialization, carbon dioxide levels remained between 200 and 300 parts per million, because of this natural balance between naturally occurring carbon dioxide emissions and carbon sinks.

Data from the National Oceanic and Atmospheric Administration (NOAA) Mauna Loa Observatory in Hawaii reflects that the concentration of carbon dioxide in the atmosphere has increased from approximately 277 parts per million (ppm) in 1750, the beginning of the industrial era, to 405.0 ppm in 2017, peaking at 414.50 in March 2020. According to Corrine LeQuéré, former director of the Tyndall Centre for Climate Change Research, most of the initial atmospheric CO_2 increase above pre-industrial levels was caused by the release of carbon to the atmosphere from deforestation and other land-use change activities. While emissions from fossil fuels started to increase during the industrial era, these carbon-based emissions only became the dominant source of anthropogenic emissions around 1950, and their relative share of total anthropogenic emissions has continued to increase until the present. It is important to keep in mind that anthropogenic emissions occur on top of a generally balanced natural carbon cycle that circulates carbon among the reservoirs of the atmosphere, hydrosphere, and lithosphere. Consequently, Dr. LeQuéré and other scientists have been leading research efforts that track not only carbon emissions, but also the effectiveness of carbon sinks, so that policymakers have the full data needed for decision making.[28]

OXYGEN CYCLE: Before leaving the carbon cycle, let's not forget that terrestrial, aquatic and marine plant life that absorb atmospheric carbon dioxide *also* release the molecular oxygen (O_2) needed for animal and human respiration. Terrestrial, aquatic, and marine plant life are essentially the lungs of the planet, taking in atmospheric carbon dioxide and releasing oxygen for terrestrial, aquatic, and marine animals to breathe. Oxygen is produced not only in our biosphere and hydrosphere by plant photosynthesis and respiration, but it is also produced in the atmosphere. In the atmosphere, molecular oxygen is created when ultraviolet radiation from the sun reacts with nitrous oxide and water vapor. The chemical reaction frees hydrogen from the water vapor and nitrogen from the nitrous oxide, with the end product being molecular oxygen.

The other side of the oxygen cycle is consumption, and oxygen is consumed by animal respiration, decomposition of organic matter, combustion, and oxidation. There are three reservoirs that hold all this oxygen: (i) the atmosphere, which is about 21% oxygen by volume, (ii) the hydrosphere, which is about 33% oxygen as a component of water and as dissolved molecular oxygen, (iii) the lithosphere, where Earth's crust and mantle hold large quantities of oxygen in solid form as silica and oxide minerals. The oxygen portion of the carbon cycle is less disrupted than some of the other systems we have discussed, so oxygen is cycling freely through its reservoirs. Still, we have to remember that when deforestation destroys our vital carbon sinks, it is at the same time eliminating one of our important sources of oxygen production. We need a functioning carbon cycle for more than combating climate change: we need it to produce oxygen for us. It's okay, go ahead and take a deep breath.

FOSSIL FUELS: We cannot discuss the carbon cycle without a few remarks about fossil fuels. The process that creates fossil fuels can be viewed as an extension of the carbon cycle, beginning at the point where terrestrial and marine plants and animals die and decay. This point is noted by a yellow arrow in the carbon cycle graphic a few pages back. Fossil fuels are buried geologic deposits of organic materials, formed from decayed terrestrial and marine plants and animals which have been exposed to heat and pressure in Earth's crust over millions of years. Below is a helpful graphic depicting the process by which fossil fuels are formed.

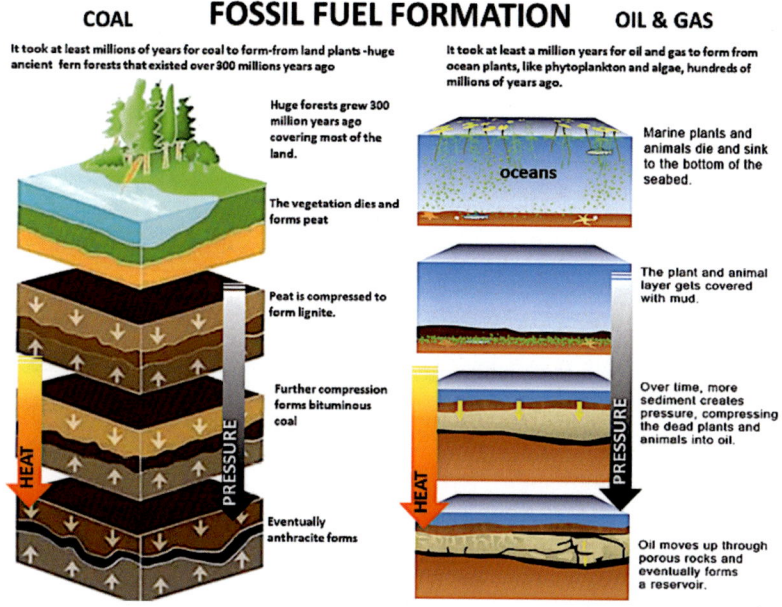

The combustion of fossil fuels has been the engine driving the industrialized world, delivering most of the world's energy needs. It has provided many of us—sadly, not all of us—with the trappings of a comfortable lifestyle. What happens when we burn fossil fuels such as coal, oil, or natural gas? Fossil fuels are molecules composed primarily of hydrocarbons, which are hydrogen and carbon atoms bonded together. Upon combustion, the carbon-hydrogen bonds are broken. Since it takes energy to hold the carbon and hydrogen atoms together, when heat causes the bonds to break, that energy is released for us to capture and use. At the same time, the now liberated carbon atoms look around the atmosphere into which they are released, hoping to meet and bond with a couple of free oxygen atoms to form carbon dioxide.

The newlywed carbon-oxygen molecules, now called carbon dioxide, will take up residence either in the atmosphere or in our oceans. If the newlywed carbon dioxide molecules head to the ocean, they will meet up with seawater, and many will get taken up by marine plant life as described in the carbon cycle. The carbon dioxide molecules not taken up by marine plant life and algae will remain in the sea and form carbonic acid, which reacts further with the seawater to form bicarbonate and hydrogen ions (In this case, hydrogen ions are hydrogen atoms that have lost their only electron in the chemical reaction with seawater). The hydrogen ions lower the pH of the water, making it more acidic. The acidic seawater is less habitable for marine plant and animal life and reduces the ability of the seawater to function as a carbon sink. The bicarbonate formed from the carbonic acid again reacts with the seawater to create even more hydrogen ions, making the seawater even more acidic.

On the other hand, if the newlywed carbon dioxide molecules take up residence in the atmosphere, they will remain married to those oxygen atoms for one to two hundred years, or longer. Only 80% of these carbon dioxide molecules will have been removed after 100 to 200 years, and the remainder can hang around for 1000 years or longer. Divorce or separation would be a good thing in this instance. Instead, the carbon dioxide molecules which have set up housekeeping in our atmosphere aggregate and make it difficult for the radiant energy from the sun to return to space after it strikes the surface of Earth. Welcome to the carbon cycle version 2.0, the one modified by the greenhouse effect.

By volume, carbon dioxide is the most abundant greenhouse gas, and it takes a very long time for it to be absorbed and removed from the atmosphere. We rely on (i) plant life, especially large forested areas, (ii) soils, and (iii) oceans to absorb all of the carbon we release. Our cutting and burning of forested areas (deforestation), overuse of once fertile croplands (desertification), and the acidification of the oceans (acidification) have together made the carbon sinks considerably less effective. At just the time we need to be removing more and more excess carbon dioxide from the atmosphere, we have reduced the functionality of our carbon sinks. As shown in the graphic on the following page, we are globally releasing increasing amounts of carbon dioxide into the atmosphere, hitting an all-time high in March 2020, a date beyond the graphic. We are reminded by Glen Peters and

Jan Korsbakken, researchers at the Center for International Climate Research, that "The continued growth in emissions simply indicates that climate policies are insufficient to overcome the continual upward march of energy use, driven both by the need to develop and the desire to consume even more."[29]

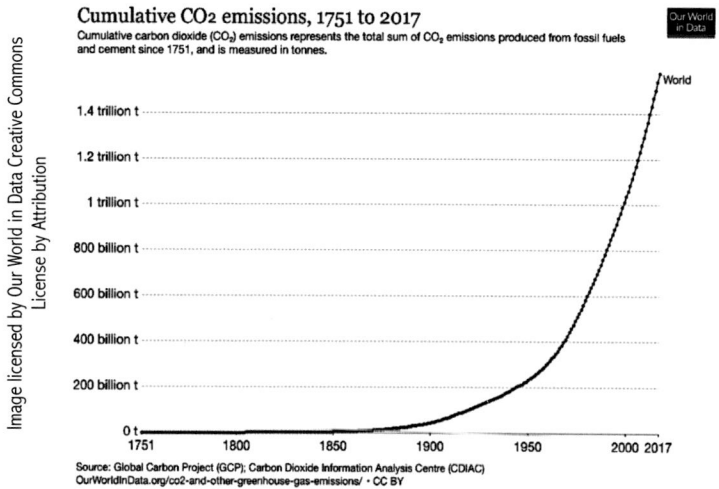

Cumulative CO2 emissions, 1751 to 2017
Cumulative carbon dioxide (CO_2) emissions represents the total sum of CO_2 emissions produced from fossil fuels and cement since 1751, and is measured in tonnes.

Source: Global Carbon Project (GCP); Carbon Dioxide Information Analysis Centre (CDIAC)
OurWorldInData.org/co2-and-other-greenhouse-gas-emissions/ · CC BY

Our appetite for fossil fuels is so insatiable that we have begun to employ even more destructive and costly technologies to extract our remaining fossil fuel reserves. I am referring to such methods as hydraulic fracturing (fracking) and tar sand mining, both of which are burdened with environmental challenges. Tar sands are sand and clay infused with a very viscous fossil fuel called bitumen. It is found in large quantities in northern parts of Alberta, Canada, a lush forested area dotted with glacier-formed lakes and wetlands. Some of these exquisite forested areas of Canada are being clear-cut for tar sand mining, destroying habitats for thousands of animals, and eliminating vast swaths of carbon sink plant life. Modeling done by the US Department of Energy Argonne Laboratory has shown that producing a barrel of crude oil from tar sands may emit as much as 20% more greenhouse gases than conventional extraction/production methods. Measurements of actual carbon dioxide atmospheric concentrations in the area of the Alberta, Canada project reflect much higher levels of emissions are taking place. Significant amounts of fresh water are used in tars sands mining, and the water is not returned to the surface or ground sources from which it came; instead, it must be impounded in "tailings ponds" because it is contaminated with toxins.

Perhaps the paradoxical "good news" is that our economically accessible fossil fuel reserves are being exhausted, as shown by the graphic below, reflecting that we have about 114 years of coal and about 50 years of natural gas and oil reserves remaining.

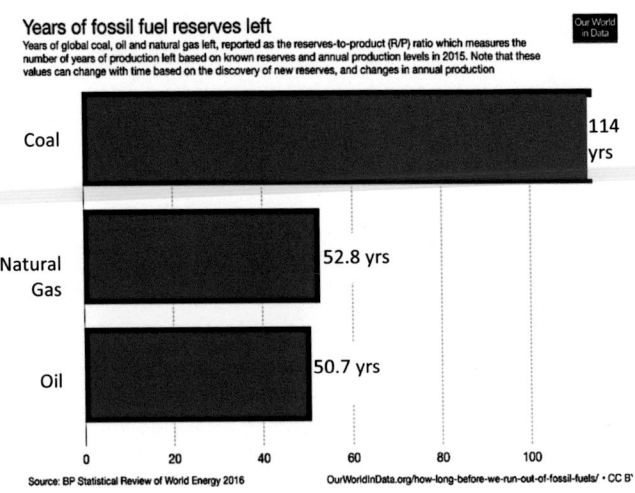

Image licensed by Our World in Data Creative Commons License by Attribution

BOTTOM LINE: FOSSIL FUELS: Carbon-based fossil fuels take millions of years to be formed. Fossil fuel combustion has been the engine of the modern world. We have a limited quantity of economically accessible fossil fuel reserves remaining, probably not enough to take us much past the end of this century. The rush is on to develop sustainable alternatives, and some of this technology is already in use. Unfortunately, we are not even close to being able to switch off the flow of fossil fuels. We will continue to emit carbon, and carbon dioxide will continue to accumulate in our atmosphere. The accumulations of carbon dioxide in our atmosphere today are a function of several factors, including (i) cumulative emissions of past centuries, (ii) current emissions from our recent activities, (iii) saturation of our carbon sinks (forests, soil, plants, and oceans) which would normally absorb the excess carbon emissions, and (iv) positive feedback loops multiplying the effects of our emissions.

Consequently, current strategies to reduce carbon emissions are unlikely to change planetary warming until sometime into the distant future, a function of the lengthy lifespan of carbon dioxide. As insidious as carbon dioxide accumulations in the atmosphere are, the complete story of global environmental sustainability only begins with carbon dioxide emissions. Remember the parable of the blind men and the elephant.

Image under license from Shutterstock.com

MECHANISM OF THE GREENHOUSE EFFECT

Sometimes I envision our Earth as a beautiful, perfectly round piece of deep blue polished lapis lazuli, encircled by veins of white calcite and pyrite matrix. The lapis represents our beautiful planet, and the matrix represents Earth's atmosphere. Earth's atmosphere is composed primarily of oxygen and nitrogen held in place by gravity. As a result of gravity, our atmosphere exerts about 14 pounds per square inch of pressure on Earth's surface. The atmosphere makes our life on Earth possible by absorbing ultraviolet radiation from the sun and protecting the surface of our planet from extremes of hot and cold as Earth spins on its axis and rotates around the sun. Without our atmosphere, our planet would be a frozen and lifeless orb making its circuits around the sun, a rather dreary affair to contemplate.

The sun provides radiant energy to Earth in the form of a continuous stream of photons. However, for Earth to maintain a stable temperature, the flow of radiant energy striking the surface of the planet must be balanced by the flow of radiant energy leaving. This balance is referred to as "Earth's energy budget." Earth's atmosphere is transparent to the shortwave radiant energy emanating from the sun, the photons from the sun bombarding us. (You can revisit the section on energy in Chapter 1 for a review of the source of these photons.) As a result of this transparency, much of the shortwave radiant energy from the sun travels through our atmosphere and strikes Earth's surface, where most of it is absorbed, warms Earth's surface, and is converted to longwave infrared energy.

Historically, Earth's atmosphere had been sufficiently transparent to this longwave infrared energy to permit the return to space of most of this energy. Regrettably, once we began releasing substantial quantities of carbon dioxide, methane, nitrous oxide, water vapor, fluorocarbons, hydrofluorocarbons, and sulfur hexafluoride gas molecules into our atmosphere, our atmosphere was no longer entirely transparent to longwave infrared energy attempting to return to space. Each type of greenhouse gas molecule blocks a distinctive amount of radiant energy trying to leave the Earth's surface. When infrared energy is unable to leave Earth's atmosphere because greenhouse gas molecules are absorbing it and re-emitting it back to Earth's surface (counter radiation), the net result is warming of the surface of our planet. This entire process is the greenhouse effect. The image below depicts the mechanisms of the greenhouse effect.

Greenhouse effect

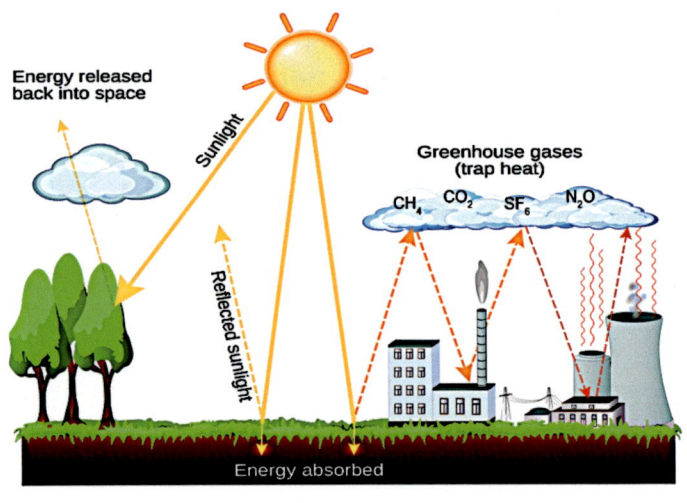

The shortwave radiation from the sun comes to us in the form of waves of visible light that can be observed during the part of the day that Earth is facing the sun. The longwave infrared radiation given off by Earth's surface is not visible to the naked eye. You read earlier that each type of greenhouse gas molecule blocks a distinctive amount of radiant energy trying to leave the surface of the planet. The amount of infrared radiant energy that a greenhouse gas molecule can block is called its *global warming potential (GWP)*. Only gases that block the return of infrared radiation are considered greenhouse gases. For instance, the atmosphere is made up a large amounts of nitrogen gas, but nitrogen gas molecules are fully transparent to infrared radiant energy trying to leave the Earth. The baseline greenhouse gas is carbon dioxide, which has a GWP of 1, with both the atmospheric lifespan and its location on the electromagnetic spectrum factored in to determine its GWP. A greenhouse gas molecule with a higher GWP than carbon dioxide can trap more heat than a carbon dioxide molecule. Methane, for example, has a GWP of 86 measured over a 20-year time span, and nitrous oxide has a GWP of 268 measured over a 20-year time span. This means that one molecule of methane is equal to 86 molecules of carbon dioxide, and one molecule of nitrous oxide is equal to 268 molecules of carbon dioxide, with regard to warming potential.

Unfortunately for those of us with a seat on spaceship Earth, GWP measures are expressed without taking into consideration the multiplying impact of the positive feedback loops caused by warming the surface of our planet. *Positive feedback loops* are added adverse effects that amplify the primary effects of greenhouse gases. For instance, with the melting of the polar sea ice, a highly reflective surface (ice) is replaced with a highly absorbent surface (seawater). Ice is very reflective, capable of reflecting into space 50% to 70% of radiant energy it receives from the sun, while the surface of seawater is a highly absorbent, resulting in a decrease in the amount of radiant energy that can be reflected back into space. Another example is the melting of permafrost due to planetary warming. Permafrost is frozen soil that has trapped and stored considerable quantities of methane and carbon dioxide over time. When the permafrost melts, it releases huge quantities of stored methane and carbon dioxide into the atmosphere, resulting in additional atmospheric greenhouse gases blanketing

Earth. One of the most cringeworthy examples of a feedback loop is the slashing and burning of tropical forests to clear land for animal feed crops and grazing. Many of these ancient hardwood trees hold centuries of stored carbon, so when the trees are burned to clear the land, the result is a virtual carbon bomb released into our atmosphere.

The warming of the surface of our planet due to greenhouse gas emissions blocking the return of radiant energy causes (i) the melting of polar sea ice and glaciers, which release their stored carbon into the atmosphere and deposit more water into our oceans, (ii) increases in the quantity of water vapor molecules in the atmosphere (water vapor is a potent greenhouse gas we will explore later), (iii) increases in the temperature of our oceans, accompanied by thermal expansion of the seawater and resulting higher sea levels, (iv) increases in precipitation quantities. The increasingly warmer air in our atmosphere is able to hold more moisture, making more water molecules available to fall to Earth as precipitation, and (v) changes in the climate patterns that drive our weather. The graphic below shows the total emissions of each of the principal greenhouse gases over the last 2000 years. Keep in mind the *global warming potential* or GWP of each of the gases when you study the graphic.

Concentrations of Greenhouse Gases 2000 Years

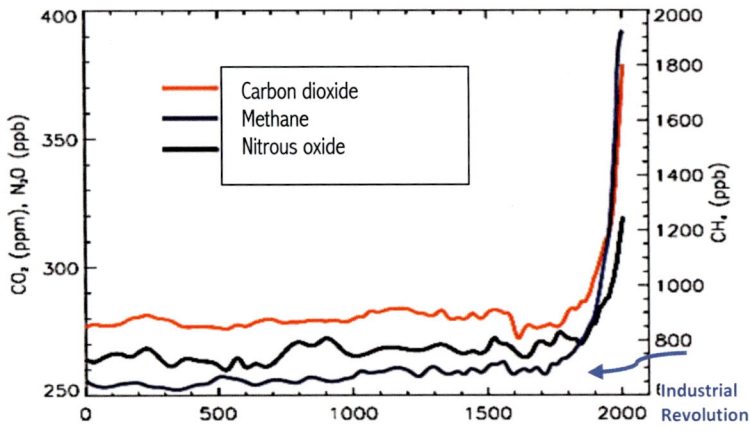

Image licensed by Our World in Data Creative Commons License by Attribution

IPCC AR4 FAQ

Land, sea, ice, the sun, and our atmosphere are the fundamental drivers of Earth's climate system, along with the rotation of Earth on its tilted axis. Ocean currents are conveyor belts of warm and cold seawater. Horizontal ocean currents send cold water at the surface of the sea from the poles toward the equator to cool the tropics and send warm water from the equator to the poles to get cooled. Vertical ocean currents, like convection in a pot of boiling water, bring cold water up from deep in the sea while sending warm water deep into the sea to be cooled. Sea currents also transfer cold or warm air to adjacent landmasses. In a similar manner, air currents high in Earth's atmosphere move hot and cold air, sending cold air from the poles to the equator, and sending warm air from the equator to the poles. The Earth's rotation on its axis prevents these moving air currents from flowing directly north or south, forcing the air currents to curve east or west (due to the Coriolis effect) to create closed air circulation loops that feed each other. Variations in warming from the sun (due to clouds and the tilt and rotation of Earth) give rise to uneven heating of Earth's surface, which in turn creates air pressure gradients. The gradients force the air masses to move from areas of high pressure to areas of low pressure, creating wind. When greenhouse gases block large quantities of solar radiation from leaving the surface of the planet, the resulting heating of the land, atmosphere, and sea prevents sea currents and air currents from providing their usual movement of heat through Earth's climate system. This change in the functioning of air and sea currents can, in turn, amplify the effects of local weather patterns; for example, by increasing the number and strength of regularly occurring hurricanes.

We know from the fossil record that Earth's climate has changed throughout history, with documented ice ages and glacier retreats. Most of those events have been attributed to small fluctuations in Earth's orbit or tilt, or events such as major volcanic eruptions or fires that could affect the amount of solar energy the planet receives. Scientists consider these phenomena to be materially different from our current warming patterns. By measuring fossilized tree rings, fossilized shells from the floor of the ocean, and dissolved gases in ice cores, scientists can attain an approximation of certain atmospheric conditions that existed thousands of years ago. For example, captured air molecules trapped in ice cores can tell us about the composition of the atmosphere ten thousand years ago. The collected data is used to chart the temperature and air composition of a given region over hundreds of years. These data consistently show that the current warming of the planet distinctly parallels the rate at which atmospheric greenhouse gases have been increasing, as well as the rate at which the human population has been increasing. See graphic from the National Oceanic and Atmospheric Administration (NOAA) below, which shows carbon dioxide emissions over the past 800,000 years, taken from ice core data.

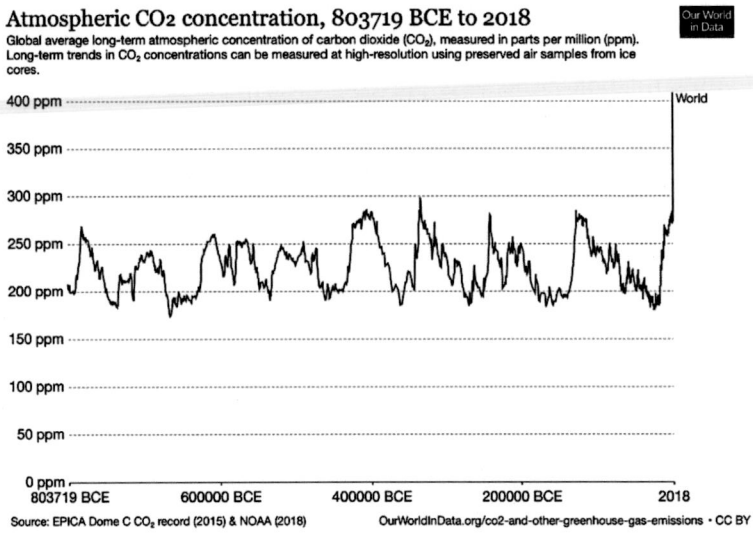

Atmospheric CO2 concentration, 803719 BCE to 2018

Global average long-term atmospheric concentration of carbon dioxide (CO₂), measured in parts per million (ppm). Long-term trends in CO₂ concentrations can be measured at high-resolution using preserved air samples from ice cores.

Source: EPICA Dome C CO₂ record (2015) & NOAA (2018)   OurWorldInData.org/co2-and-other-greenhouse-gas-emissions · CC BY

From NOAA, provided by Nat Ctr for Environ Inform, per terms of use
https://www.climate.gov/news-features/understanding-climate/climate-change-atmospheric-carbon-dioxide

The Intergovernmental Panel on Climate Change (IPCC) is an independent intergovernmental body of the United Nations created in 1988 and currently has 195 member countries. The IPCC is tasked with providing governments and policymakers with scientific information that they can use to develop climate policies. The data and conclusions of the IPCC are open to review and fully available on the IPCC website. This organization has become the world's leading and accepted authority on climate change. Its reports have been supported by the world's leading climate scientists and the governments of its 195 constituent countries. The IPCC has unequivocally and repeatedly stated that global warming is taking place at an increasing rate and that most of the warming is a result of greenhouse gas emissions created by the activities of humans. I urge you to visit the IPCC website at https://www.ipcc.ch to read the reports and findings of this organization.

At the direction of the United Nations, the IPCC issued a special report in November 2018 advising that the climate actions pledged by nations of the 2015 Paris Climate Agreement are inadequate to meet the 1.5°C to 2°C agreed maximum increase in greenhouse gas emissions. They concluded that we may have as little as 12 years [until 2030] to reach the 1.5°C target, and it will take stronger actions than the actions pledged by participating countries to achieve this target. They also concluded that 2°C of warming is materially more damaging than the 1°C increase we have experienced since 1970. The report concludes that current global emissions place us on a trajectory toward 3°C warming by 2100.

In this same report, the IPCC attempted to quantify the costs of the damage caused by planetary warming. The IPCC estimated global economic damages of $54 *trillion* by 2100 if the planet warms by 1.5°C and $69 *trillion* by 2100 if the planet warms 2°C. A report by the National Resources Defense Council expresses losses for the United States in terms of US gross domestic product, concluding that we should expect as much as 3.6% of GDP *per year* in damage from more intense hurricanes, real estate losses from sea level rise, increased energy costs from excess cooling needs and reductions in hydroelectric power due to drought, and agricultural losses from water scarcity.[30] "When global warming takes hold, there could be as many as 200 million people overtaken by disruptions of monsoon systems and other rainfall regimes, by droughts of unprecedented severity and duration, and by sea-level rise and coastal flooding."[31] Whoever thought that we would have to add the term "environmental refugee" to our lexicon? However intimidating the costs of mitigation may be, mitigation costs have to be lower than the costs of replacing or repairing damaged infrastructure and property. Could it be that future global economic well-being depends on the execution of extensive mitigation technologies?

Methane emissions have become a significant issue. Methane is used as fuel (a principal component of natural gas) and as part of various industrial processes, but it is significant primarily because of the magnitude of harm it can do to our environment. As you can see from the graphic on the following page, methane cycles through Earth's ecosystems by way of various methane sources and methane sinks. Most methane is created by microbes called methanogens in a totally anaerobic (without oxygen) process called methanogenesis. It occurs in the soil, fresh water, and marine environments. Methanogens are strict anaerobes and biologically constrained to produce methane; they cannot function in the presence of oxygen. As with atmospheric carbon dioxide, the amount of atmospheric methane produced by natural sources was historically kept in check by methane sinks. Atmospheric methane concentrations have increased by 250% since the beginning of the industrial period (e.g., the mid-1700s). The inability of the methane cycle to maintain a stable level of atmospheric methane means that the methane cycle, like the carbon cycle, is disrupted. Can you guess who broke it? The sources that generate methane and the sinks that consume methane are described below.

METHANE SOURCES: The primary sources of atmospheric methane, anthropogenic and natural, are as follows:

- Natural decay of organic materials (e.g., microbes in the soil and wetlands)
- Emissions released from Earth's crust during oil and gas exploration
- Leaks in the transmission and storage of natural gas
- Emissions from the digestive process of termites (who knew?)
- Biomass burning, including forest fires, charcoal combustion, and firewood
- Rice paddy agriculture (excess fertilizer, flooding of paddies and decay)
- Livestock farming (manure management and ruminant enteric fermentation)

- Decay of organic materials in landfills
- Municipal wastewater treatment
- Melting permafrost and polar ice (both are former methane sinks now emitting centuries of stored methane as they melt)
- Emissions from methane seeps or vents in the seafloor. Methane formed from buried organic carbon that has dropped to the seafloor is trapped in the rock below the seafloor. When methane bubbles up through these seeps, it reacts with the sea water and can work its way to the top of the water column to be released as carbon dioxide.

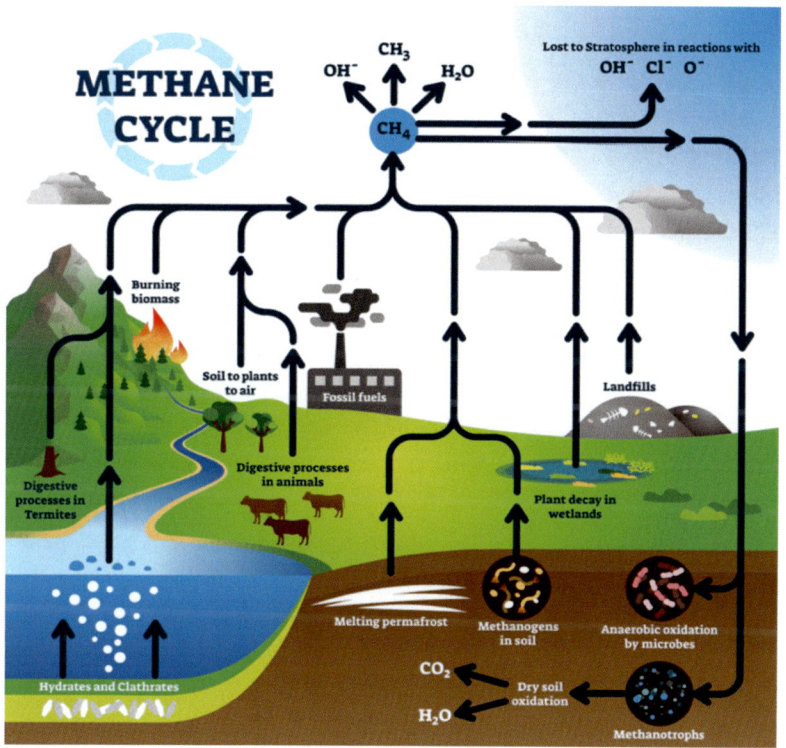

Image under license from Shutterstock.com

Note that most of the sources shown in the graphic are from human activities, and since the 1980s, emissions of methane from human activities have been almost double the emissions from natural sources.

METHANE SINKS: The primary natural process of removing methane from Earth's atmosphere is the oxidation of the methane by hydroxyl radicals high in the atmosphere. A hydroxyl radical is a molecule made up of one oxygen atom bonded to one hydrogen atom, and it is formed high in the troposphere when water vapor molecules come into contact with oxygen molecules. Hydroxyl radicals are mighty methane sinks. They behave like little Pac-Men, gobbling up methane molecules from the atmosphere and breaking them down into water and a very stable gaseous CH_3. Another robust methane sink are the single-celled bacteria known as methanotrophs, found in wetlands, landfills, rice paddies, and the soils of forest floors. Methanotrophs consume atmospheric methane for energy. In April 2020, a paper published in the journal *Science Advances* described a newly discovered methanotroph found near the openings of deep ocean seeps. The researchers discovered that a certain species of worm typically located near these seeps had a symbiotic relationship with a particular species of methane-consuming bacteria, allowing the worms to assimilate methane through their skins when the bacteria attach themselves to the worms. Presto, an unknow seaworm becomes a methane sink![32] Because more methane is being released into the atmosphere than can be absorbed by our natural hydroxyl sinks and our hard-working methanotrophs, atmospheric methane concentrations are increasing.

So, the good news is that, unlike carbon dioxide, which stays around in Earth's atmosphere for 100 to 1,000 years, methane is absorbed in about 10 to 15 years. The bad news is that the blanketing and warming effect of methane in our atmosphere is orders of magnitude greater than that of carbon dioxide because methane traps 86 times more heat over a 20-year cycle than carbon dioxide. We referred in the section on greenhouse gases to this measure of the blanketing and warming effect of a greenhouse gas as GWP. You will also encounter the term "radiative forcing," which, in this context, is the net change in the energy balance of Earth due to the specific greenhouse gas. It measures the relative influence a greenhouse gas has in altering the balance of incoming and outgoing thermal energy in Earth's atmosphere. Turn back to the greenhouse gas graphic if you need a refresher.

ATMOSPHERIC METHANE: For over 50 years, NOAA has collected air samples from all over the globe as part of one of the world's longest-running greenhouse gas monitoring programs. The website link is https://www.esrl.noaa.gov/gmd/ccgg/. The data show that concentrations of methane in the atmosphere have steadily increased, but then leveled off in the year 2000. In 2007, methane concentrations abruptly started rising again. It appears that the scientific community did not understand the reason for this and did not expect this abrupt change. Suspects include hydraulic fracturing (fracking), increased numbers of livestock, the loss of traditional sinks to absorb excess methane, and declining levels of hydroxyl radicals, which are engaged in neutralizing other anthropogenic pollutants. Another suspect is our natural gas line system. Methane is a primary component of natural gas, which travels in gas lines to our homes, schools, and businesses.

In June 2018, a research paper was published in the journal *Science* demonstrating that US oil and gas operations are leaking 60% more methane gas than the US government EPA estimates had predicted.[33] More current research reported in *Nature* magazine in early 2020 has concluded that atmospheric emissions of fossil methane, such as those that seep out of vents deep in the sea, make up a much smaller fraction of total methane emissions than previously thought. The volume of methane emitted from the seafloor is significant, but the amount of methane that makes it to the surface of the sea is much less, in part due to the short life of methane in seawater.[34] Research published in 2016 in the journal *Global Biogeochemical Cycles* suggests that anthropogenic microbial sources are the most prevalent sources of current methane emissions. "Evidence from carbon isotopes implies that the primary cause of the new growth is an increase in biogenic emissions, probably from wetlands and also agricultural sources, such as rice fields and cattle." Other Anthropogenic microbial sources would include emissions from manure management systems, landfills, and other waste decaying processes initiated by humankind.[35] The growing livestock agriculture sector, driven by increasing demand for meat, could account for much of this. In Chapter 5, I cite research reflecting that as much as 37% of our total methane emissions may be from the livestock agriculture sector.

The share of methane emissions attributable to anthropogenic activity versus natural processes will continue to be a topic of debate until we have better data to confirm both. In 2021, the United States will be launching a satellite (MethaneSat) that will provide global, high-resolution coverage of methane emissions from oil and gas facilities, agriculture, and other sources. It will identify the location of methane emissions and the amount of methane being released. The project is being privately funded and is sponsored by the Environmental Defense Fund (EDF). Steven Hamburg, Chief Scientist for MethaneSat, describes the project as follows on the EDF website: "It's going to greatly increase our ability to monitor methane emissions from human activities on a global scale. The notion that we can see greenhouse gas emissions at a scale that's relevant to effective mitigation can be a game changer for overall greenhouse gas reductions."

Russia and the countries of the former Soviet Union produce significant quantities of natural gas. Some researchers suspect that after the breakup of the Soviet Union and the loss of its centralized government, the aging oil lines and related equipment across the former Soviet states continued deteriorating and are now releasing large amounts of methane. There is so much methane leaking over the former Soviet states that some researchers have suggested that just upgrading the gas lines of the former Soviet Union could have a significant impact in reducing our global methane emissions.

BOTTOM LINE: METHANE: Any way you look at methane, you have to conclude that while it may not be the most abundant greenhouse gas in the atmosphere, it is definitely among the most robust—and by robust, I mean harmful. There is no question that methane plays a leading role in greenhouse gas accumulations and planetary warming, owing to its high GWP. The optimist's view of this statement is that if methane is so harmful as a greenhouse gas, why don't we shift a portion of our near-term focus from carbon dioxide to methane? This notion becomes even more relevant if we learn from MethaneSat exactly where and how much methane is being released. This shift might allow us to achieve more immediate results in reducing global warming and might keep us from the 1.5° to 2°C tipping point. The shift also might buy us time to work on the larger (by volume) carbon dioxide problem. We should grab any low-hanging fruit we can find.

NITROUS OXIDE

Yes, laughing gas; however, nitrous oxide (N_2O) in the atmosphere is not a laughing matter, and it is not a result of too many dentists performing too many root canals. As reflected in the graphic of the nitrogen cycle in Chapter 2, nitrous oxide is produced naturally from microbial action in the soil, melting permafrost, and our oceans. It is also produced by human activity through our use of commercial fertilizers, wastewater treatment methods, livestock manure management, fossil fuel combustion, and some chemical industrial processes. According to the US Environmental Protection Agency (EPA), approximately 40% of total nitrous oxide emissions are anthropogenic, and approximately 75% of anthropogenic nitrous oxide emissions are from agriculture. Agricultural sources of nitrous oxide include (i) manure lagoons that emit nitrous oxide from manure at the bottom of the lagoon where there is little oxygen, (ii) organic fertilizer, aka manure, and (iii) overuse of synthetic nitrogen fertilizer.[36] While the nitrogen in synthetic fertilizer is in a chemical form that can be immediately absorbed by plants, most of the naturally occurring nitrogen must undergo chemical changes before it can be used by plants. During this conversion period, large quantities are lost to the atmosphere as nitrous oxide. Compared with carbon dioxide, the amount of nitrous oxide we are releasing into the atmosphere is small (6% to 8%). Even so, nitrous oxide is considered one of the top three greenhouse gases because its warming potential is 268 times that of carbon dioxide. Stated another way, one molecule of nitrous oxide is equivalent to 268 molecules of carbon dioxide, and it hangs around in the atmosphere for about 120 years.

When nitrous oxide resides in the lower portions of our atmosphere, it remains relatively inert and operates as a potent greenhouse gas. However, as the nitrous oxide (N_2O) molecules float upward to the stratosphere, they are exposed to sunlight and oxygen and converted to nitric oxide (NO). Nitric oxide then reacts with the ozone molecules (O_3) that form the protective ozone layer that keeps Earth from receiving too much ultraviolet radiation from the sun. This reaction depletes ozone molecules, reducing our protection from ultraviolet radiation. You have probably heard about chlorofluorocarbons (CFCs), which made headlines in the 1980s when researchers discovered that CFCs were damaging the ozone layer. There were breaks in this protective layer at both poles. The international community gathered in Montreal in 1987 to address the depletion of our ozone layer. The result was the Montreal Protocol, which set forth regulations for the use of CFCs and specific other ozone-damaging gases. Since 1987, the presence of atmospheric CFCs has been greatly reduced. Unfortunately, nitrous oxide was not included in the restrictions of the Montreal Protocol, and it is currently the most critical threat to our ozone layer.

In light of the importance of food production in a world where hundreds of thousands of people are nutritionally deprived or outright starving, reactive nitrogen and nitrous oxide mitigation methods have been a challenging problem to tackle. There is much research taking place to define and test "best practices" for nitrogen fertilization in crops. Practices such as cover crops, reduced use of fertilizer, selection of low nitrogen-demand crops, different irrigation methods, optimal timing of fertilizer application, and reduced tilling of the soil all show promise in research.[37]

BOTTOM LINE: NITROUS OXIDE. Nitrous oxide is another robust greenhouse gas, with some researchers calling nitrous oxide the next carbon dioxide. Yet, we hear not a peep from policymakers about nitrous oxide. As I mentioned in the introduction, our issues are much greater than our collective carbon footprints.

WATER VAPOR

Water vapor is the gaseous step of the water cycle, and it is a potent greenhouse gas. Interestingly, water vapor is the most abundant of the greenhouse gases, but we rarely discuss water vapor when considering anthropogenic climate change. The principal reason is that water vapor has a short cycle in the atmosphere, approximately 10 days, before it is incorporated into weather events and falls to Earth. Consequently, it does not build up in the atmosphere in the same way that carbon dioxide, methane, and nitrous oxide do. However, water vapor can create an alarming feedback loop, as demonstrated by the following chronology:

- Carbon dioxide, methane, and other greenhouse gases are emitted into the atmosphere, where they take up residence because our sinks are inadequate to absorb all of the emissions.

- The greenhouse gases residing in the atmosphere prevent the sun's solar radiation reflecting off Earth's surface from returning to space, causing Earth's air temperatures to increase.

- Due to warmer air temperature, more of Earth's surface water evaporates into the atmosphere and becomes water vapor.

- The warmer atmosphere is able to absorb more water molecules of water vapor than before, resulting in an increase in the volume of atmospheric water vapor. There is a maximum amount of water molecules our atmosphere can retain at any given temperature. Relative humidity reflects the percentage of water saturation of the atmosphere at a given time.

- The additional water vapor, like the other greenhouse gases, blocks the departure of even more solar energy reflected off Earth's surface, causing the temperature at the surface of Earth and lower atmosphere to rise even further. Compounding this is the fact that water vapor can block and absorb a broader range of wavelengths of radiant energy than the other greenhouse gases.

- The atmosphere, like a sponge, becomes saturated and cannot retain any more water vapor molecules, so it the vapor condenses and is released as precipitation.

- And the cycle starts over, with an atmosphere that is even warmer and can retain even more water vapor molecules.

Interesting research headed by Andrew Dessler from Texas A&M has demonstrated that the water vapor positive feedback loop at the lowest level of the atmosphere, the troposphere, approximately *doubles* the initial warming caused by carbon dioxide. For example, if atmospheric carbon dioxide concentrations double, there is usually a proportional 1.2°C average increase in global temperature. However, when the atmosphere has warmed and is holding greater quantities of water vapor molecules, a doubling of atmospheric carbon dioxide will cause a 2° to 4.5°C increase in temperature. Dr. Dessler has concluded that a significant portion of the projected warming of Earth is from feedback loops which amplify the initial warming effect. More recent research by Dr. Dessler is pointing to water vapor in the next layer up from the troposphere, the stratosphere, as implicated in a second water vapor feedback loop. As one of the researchers in the Texas A&M study of water vapor stated, "yet another reason to hate humidity." (Clearly, this is a reference to the empirically well-established and troubling direct correlation between humidity and bad hair days).[38]

Chapter 4

POPULATION IMPACTS

POPULATION AND SUSTAINABILITY

Although population is a significant factor in defining the carrying capacity of our planet and achieving sustainability, this subject has been conspicuously absent from the public sustainability discussion. Remember Thomas Malthus from Chapter 1? Malthus was writing his essay on population during the Age of Enlightenment and the beginning of the Industrial Revolution. It was a time of unimagined intellectual discoveries, the spread of knowledge through improvements to the printing press, an emphasis on scientific inquiry, and developments in political philosophy that illuminated the Western world with hope. The Western world was imbued with the belief that it was on a trajectory of continuous improvement, and that population growth and economic growth could continue ad infinitum. Malthus agreed that new technologies would likely increase the efficiency of food production and allow the well-being of populations to increase; yet, he firmly believed that this improvement would be short-lived. He did not think the technological improvements could sustainably provide relief to population growth, because each new level of well-being would inevitably lead to more population growth, which in turn would restore the original per capita level of food availability.

The heart of Malthus's argument was that population tends to grow geometrically, while food production (and arable land) can only grow arithmetically. He did concede that disease, war, and famine do provide "checks" on this oscillation; however, he maintained that the advancement of humankind would, nonetheless, continue to oscillate between plenty and scarcity, well-being and struggle. He believed that Earth's carrying capacity would demand, one way or another, that we keep our population within numbers that can be supported by existing resources.[39]

In 1992, the Union of Concerned Scientists (UCS) issued a warning about population growth, endorsed by 1,700 independent scientists. The UCS is a nonprofit science advocacy organization founded in 1969 at Massachusetts Institute of Technology. It was formed as a vehicle to examine US government policy in areas of science and technology and to ensure that critical environmental and social problems remain the objectives of such policy. Their comments on population growth are as follows:

> Pressures resulting from unrestrained population growth put demands on the natural world that can overwhelm any efforts to achieve a sustainable future. If we are to halt the destruction of our environment, we must accept limits to that growth. A World Bank estimate indicates that world population will not stabilize at less than 12.4 billion, while the United Nations concludes that the eventual total could reach 14 billion, a near tripling of today's [sic 1992] 5.4 billion. But, even at this moment, one person in five lives in absolute poverty without enough to eat, and one in ten suffers serious malnutrition.[40]

In 2017, the same group of scientists issued a second warning, now with nearly 15,000 signatories. While the 2017 warning reiterated many of the points of the 1992 warning, it also advised that we are approaching the point where the harm we are doing is irreversible.

Well, here you are today, possibly reading this book on an electronic device, your stomach is likely full, and you are sheltered comfortably from the elements. On spaceship Earth, the global population has grown from one billion when Malthus was writing to 7.7 billion today in early 2020. At the same time, most of the 7.7 billion residents of this planet enjoy levels of technology, convenience, comfort, sanitation, leisure, food, choices, and overall plentitude unimaginable in the history of humankind. Was Malthus wrong? One may argue that many of the natural checks on population described by Malthus have been overcome by developments in medicine, efficient food production, sanitation, technology, and relative peace since World War II (well, somewhat relative peace).

As a species, we *Homo sapiens* are heady with our accomplishments and have been behaving as though an unlimited population can be supported by the finite resources of this planet. Are we so arrogant about our technology that we believe that technological improvements can keep up with current population growth without impacting the finite natural resources upon which we rely? Or is the urge to procreate so profoundly embedded from millions of years of evolution that we can't save ourselves? What is it about the human psyche or the deeply held values and conventions of our social groups that drive us to turn a blind eye to this issue, decade after decade? Perhaps the reality of a planet without plentitude is so abstract that our brains simply cannot grasp the idea sufficiently to take action that will actually make a difference. I do not know the answer, but I do know that history is filled with examples of people turning a blind eye to severe, even life-threatening threats—always, always to their peril.

Discussions about the relationship between population and environmental sustainability are fraught with political and racial overtones that tend to make this topic one of obligatory omission— virtually taboo. As soon as you suggest that population growth should be part of the sustainability discussion, many critics will immediately link your suggestion to the eugenics practices of the Nazi Party in World War II Germany and laws in the United States in the early 1900s permitting forced sterilization of persons of color and persons with "undesirable traits." The one-child policy in China was adopted in response to overpopulation concerns of the Communist leaders, but, in practice, it became coercive and draconian with involuntary sterilizations, forced abortions, and intimidation. All such practices were despicable, racist, and inhumane, and were intended to reduce the political and economic power of targeted population groups. So, the influential voices of our policymakers continue to avoid any discussion of population in the context of environmental issues. After reviewing hundreds of carbon-reduction measures described in high school textbooks and government documents from Canada, the United States, and the European Union, a research team exploring the carbon legacy of childbearing found *not one* mention of the notion that bearing fewer children might be another vehicle for mitigating carbon emissions. [41] In the 1970s, overpopulation was considered a critical environmental issue, discussed openly and urgently. We must find a way to make this a politically safe topic of discussion once again.

Quantifying the relationship between population and environmental sustainability should arguably lessen any underlying bias by making it more objective. In the 1970s, (yes, I know, 50 years ago), Paul Ehrlich, a biologist at Stanford University, developed the following formula to describe the relationship between population and environmental impact as follows: *Impact = Population x Affluence x Technology*. Although Ehrlich was widely criticized for many of his comments and predictions, his model continues to be used by researchers today to analyze anthropogenic environmental impacts. It is useful both as a qualitative tool and a quantitative tool. *Environmental impact (I)* consists of resource depletion and waste accumulation and is considered the dependent variable in this equation. *Population (P)* refers to the size of the human population. *Affluence (A)* refers to economic consumption. To quantify this variable, per capita gross domestic product is used as a proxy for consumption. Technology (T) refers to the technical processes used to obtain resources and transform them into useful goods. It is quantified by various proxies, such as amounts of greenhouse gas emissions, the extent of nonrenewable extractive resources consumed, or global technology patents issued. Ehrlich concluded that an increase in any one of P, A, or T would increase the adverse environmental impact, but an increase in more than one of these variables would have a compounding impact on the environment. Underlying Ehrlich's model is the notion that the biosphere is made up of regenerating biogeochemical systems that can endure only if the population does not overload and disrupt the functioning of the systems.[42]

In 1972, a team of 16 researchers at the Massachusetts Institute of Technology led by Donella (Dana) and Dennis Meadows came together to research worldwide population and economic growth. They examined multiple factors: population demographic patterns, agricultural output, resource depletion, industrial output, and waste generation. They developed a statistical model to determine probable outcomes, using multiple sets of assumptions. I am envisioning an enormous, unimaginably slow computer and punch cards. When the results finally spilled forth, they concluded, all those years ago, that our planet could not support the then-current rates of economic and population growth much past the early 21st century, even with advanced technology. These findings were published in 1972 in a book entitled *The Limits to Growth.* The book was updated in 2004 and recast in 2012 by members of the team.[43]

More recently, using multinational data from several decades, a study by a research group headed by Brian O'Neill of the University of Denver revealed that after controlling for other variables, a 1% increase in population is generally associated with a 1% increase in carbon emissions. O'Neill went on to write an essay on this subject, published in the esteemed journal *Science* in 2018. He has urged IPPC and other policymakers to recognize that population growth as a critical aspect of environmental sustainability. He acknowledges that some policymakers have finally recognized the relationship between environmental sustainability and population growth, but criticizes them for not doing enough to promote public policy to address this critical issue.[44] For example, world leaders at the United Nations Conference on the Environment and Development (UNCED) held in Rio de Janeiro in 1992 drafted a declaration that emphasized the need for a two-pronged approach specifically, (i) sustainable patterns of consumption and production and (ii) policies that address managed population growth. "To achieve sustainable development and a higher quality of life for all people, states should reduce and eliminate unsustainable patterns of production and consumption and promote appropriate demographic policies."[45] The vagueness of the phrase "appropriate demographic policies" leaves one searching for detail, detail that is only now, almost two decades later, starting to emerge. Admittedly, part of the challenge of addressing high population growth rates at the policy level is that the highest fertility rates are occurring in the least developed areas of the planet that have limited infrastructure for implementing change. Additionally, there is the awkward fact that the fertility rates and consumption patterns of the most developed areas of the planet have brought us to this level of environmental unsustainability. How do we now have the temerity to ask populations in the least developed areas to reduce birthrates? In addition, population control is a provocative subject in both the developed and the developing world, opposed by conservative groups and religious leaders in both areas. To address family planning in less-developed countries is an enormous, but surmountable, challenge.

Another research team led by Brian O'Neill addressed this concern in a paper published in the journal *Proceedings of the National Academy of Sciences*. The team concluded that "Substantial changes in population size, age structure, and urbanization are expected in many parts of the world this century. Although such changes can affect energy use and greenhouse gas emissions, emissions scenario analyses have either left them out or treated them in a fragmentary or overly simplified manner."[46] This paper concludes that until greenhouse gas modeling takes into consideration all of the various factors of population dynamics, our greenhouse gas emission projections will be inaccurate. Research from this group concludes that an aging population, such as Japan's, can reduce carbon emissions by up to 20%, while migration to megacities can increase emissions by more than 24%. The group concludes as follows: "Using an energy–economic growth model that accounts for a range of demographic dynamics, we show that slowing population growth could provide 16% to 29% of the emissions reductions suggested to be necessary by 2050 to avoid dangerous climate change. We also find that aging and urbanization can substantially influence emissions in particular world regions."[47] By the end of this chapter, you will almost certainly concur that modern population patterns and population growth momentum are significant components of our metaphorical elephant in the room and that we ignore such population trends at our peril.

It might be helpful to look back before we look forward, to have a perspective of how we came to have the population we have today. It is believed that the first evolutionary descendants of modern man emerged around 300,000 to 200,000 years ago in East Africa; although, some evolutionary biologists cite evidence supporting the emergence of *Homo sapiens* much earlier in time. Early man was a hunter and gatherer, a forager who spent most of his day in small groups in search of plants, small animals, birds, and insects to eat. He developed some fundamental tools and weapons and was then able to hunt larger animals, but tended toward finding the most calorie-rich food he could obtain for the least effort. (Sound vaguely familiar?)

The climate during these years was not particularly hospitable or stable, keeping early man on the move seeking a more favorable environment. When he encountered other groups, the groups would exchange information, and sometimes the groups would combine or otherwise intermingle. He traveled light, and the nomadic lifestyle was not conducive to rapid population growth. About 100,000 to 60,000 years ago, *Homo sapiens* began leaving the African continent. About 15,000 years ago, the planet began cooling with another ice age, and *Homo sapiens,* ever the snowbird, headed south to warmer climates until the end of that last ice age, when he then populated much of the remainder of the planet. About 12,000 years ago, our planet reached a goldilocks zone, where the climate became more stable, predictable, and favorable for humanity to thrive.

About 11,000 years ago, a revolutionary era for *Homo sapiens* began when he started growing plants for food and began herding and breeding animals for food. Once he gave up the nomadic life, *Homo sapiens* was able to remain in a single location, to refine the quality of the crops he was growing, store surplus crops, build shelter from the elements, develop a division of labor and specialized skills in the group, and assemble into villages and small towns. With more food available, populations began to increase. The transition to agriculture appears to have begun in the Fertile Crescent, a crescent-shaped area starting at the very northeastern edge of Egypt, and including a swath of land going all the way to the Persian Gulf; although, there is evidence suggesting that the transition could have been happening in many different areas and at different times. Wherever and whenever such agricultural transition took place, it resulted in increased population growth. The agricultural transition also gave rise to particular adverse consequences that limited population growth. For instance, agriculture led to a conversion to cereal-based diets, which may have reduced life expectancy and resulted in illnesses from nutritional deficiencies. Similarly, the transition to a non-nomadic life with rudimentary sanitation practices allowed for disease and infections to move through these now larger groups with ease.

In 10,000 BCE, the world population was estimated to be about four million, growing to nearly 200 million by the birth of Christ. By 1803, Malthus's time, the world population reached the first billion. By 1927, the population had doubled to 2 billion. In 1960, the planet held 3 billion people. By 1975, we had added another billion, for a total of 4 billion. Each of the successive additional billions took approximately 12 years, and today in 2020, we are on the cusp of 8 billion. As you can see from the graphic on the following page, almost all of the population growth has taken place very recently, starting around the time of the Industrial Revolution. It is almost impossible to fathom that there were periods in our distant history when *Homo sapiens* was faced with extinction!

The average global population growth rate is currently around 1.05% per year as of 2019. There is no question that our average population growth rate has been globally slowing since the 1960s when it peaked at 2.2% per year. Even with historically low population growth rates, why does it sometimes feel like our planet is overrun with humans? Various population dynamics discussed in the next section will explain why our population is continuing to increase, impacting resource depletion and degradation, and disrupting our natural biogeochemical cycles.

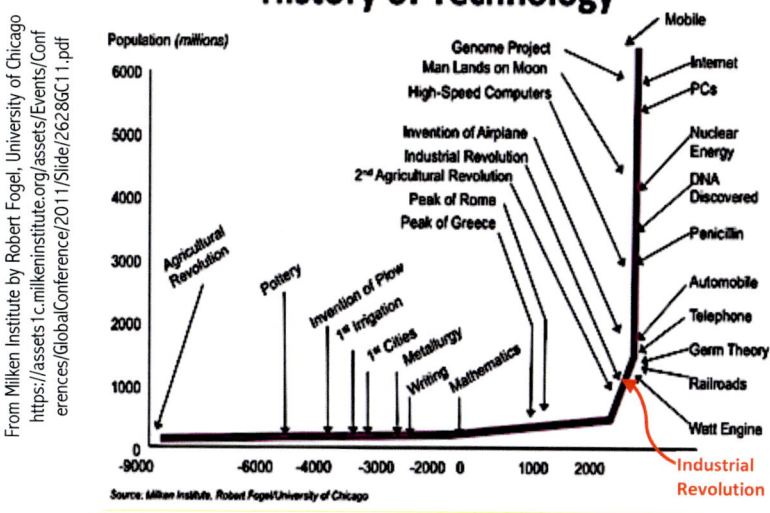

Growth of World Population and the History of Technology

Source: Milken Institute, Robert Fogel/University of Chicago

POPULATION DYNAMICS

While our global population is growing at slower rates than in our recent past, several clear population trends have been identified that explain why our planet continues to feel so crowded. The current demographic trends identified by the United Nations in its 2019 population report are as follows: (i) world population is continuing to grow, but at a slowing rate; (ii) some countries are facing very rapid population growth as a result of high fertility rates and a disproportionately large youth population; (iii) other countries are experiencing decreasing population, due to low fertility rates and/or emigration; (iv) the global population is experiencing unprecedented aging, where a disproportionate share of the population is older than 60; and (v) global migration has become an important factor in population dynamics in some parts of the world.[48] Citing the O'Neill research again, changes in population composition can have a significant influence on greenhouse gas emissions in particular regions, separate from the effect of changes in population size. "Aging can reduce emissions in the long term by up to 20%, particularly in industrialized regions. In contrast, urbanization can lead to an increase in projected emissions by more than 25%, particularly in developing regions…" O'Neill attributes both of these conclusions to the impacts that the aging and the urban populations have on labor productivity.[49]

PLAIN OLD POPULATION GROWTH: The population of our planet continues to grow, but it is growing at one of the lowest rates since World War II. Yet, in the next 30 minutes, almost 4500 babies will be born, as we continue to add 220,000 humans per day and 80,000,000 humans per year to the planet. Yes, worldwide fertility rates are nearly half of what they were 50 years ago. Yet, globally, we are living longer than at any time in history, enjoying greatly reduced infant and child mortality rates, and in certain regions of the world, there exists a disproportionately large number of young women of childbearing age. Today, there are twice as many women of childbearing age as in 1960, there are two births for every death, and almost 40% of the world's population is age 24 or younger.[50]

With these factors, it is not difficult to understand why our world continues to be very crowded. These factors have together created *population growth momentum* that cannot be stopped overnight. This is evidenced by the projections shown by the dotted lines on the graphic on the following page from the United Nations Department of Economic and Social Affairs, Population Division (UN-Population Division). The variants are based on different average fertility rates. With current fertility rates, our global population could be as high as 26 *billion* by the end of this century; although, with fertility rates continuing to drop, the medium variant projection for the end of the century is about 11.5 *billion*. It may surprise you to learn that population growth momentum can extend out as far as 70 years past the time that replacement level fertility is reached (replacement fertility is approximately 2.1 children per woman).

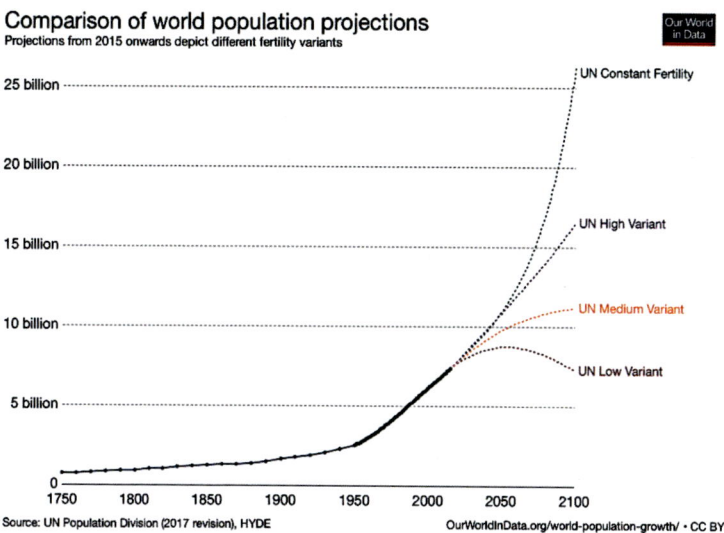

Comparison of world population projections
Projections from 2015 onwards depict different fertility variants

UN Constant Fertility

25 billion

20 billion

UN High Variant

15 billion

UN Medium Variant

10 billion

UN Low Variant

5 billion

0
1750   1800   1850   1900   1950   2000   2050   2100

Source: UN Population Division (2017 revision), HYDE

OurWorldInData.org/world-population-growth/ · CC BY

There appears to be no dispute that slowing population growth will make a meaningful impact on the health of our planet, particularly carbon dioxide emissions. I will repeat a portion of an earlier comment from the O'Neill research group: "We show that slowing population growth could provide 16% to 29% of the emissions reductions suggested to be necessary by 2050 to avoid dangerous climate change." [51] According to the Worldwatch Institute, a nonprofit environmental think tank, the overriding challenges facing our global civilization are to curtail *climate change* and *slow population growth*. "Success on these two fronts would make other challenges, such as reversing the deforestation of Earth, stabilizing water tables, and protecting plant and animal diversity, much more manageable. *If we cannot stabilize climate and we cannot stabilize population, there is not an ecocycle on Earth that we can save* [emphasis added]."[52] Think back to the recycling natural biogeochemical cycles we discussed in previous chapters—these cycles are the life support systems for all of us here on spaceship Earth, and we have disrupted nearly every one of them.

The carbon legacy component of childbearing is a critical issue explored in depth by Paul Murtaugh and Michael Schlax, both of Oregon State University, in an article entitled "Reproduction and the

Carbon Legacies of Individuals." They state, "Our basic premise is that a person is responsible for the carbon emissions of his descendants … for example, a mother and father are each responsible for one half the emissions of their offspring, and one-fourth of the emissions of their grandchildren." They concluded from their research that the "carbon legacy" of just one child can produce 20 times more greenhouse gases than a person can save by driving a high-mileage car, recycling, giving up meat consumption, using energy-efficient appliances and light bulbs, etc." [53] Murtaugh and Schlax conclude, "Future growth amplifies the consequences of people's reproductive choices today, the same way that compound interest amplifies a bank balance . . . Clearly, the potential savings from reduced reproduction are huge compared to the savings that can be achieved by changes in lifestyle."[54]

More current research headed up by Seth Wynes and Kimberly Nicholas of the University of British Columbia, reflected by the graphic on the following page, concluded that the carbon dioxide emissions reduction of having one fewer child equated to a reduction of 58 tons of carbon dioxide for each year of a parent's life. They determined this amount by adding the emissions of the child and all its descendants, then dividing this total by the parent's lifespan. Each parent was ascribed 50% of the child's emissions, 25% of their grandchildren's emissions, and so on, using the formulation developed by Murtaugh et al. cited above.[55] For perspective, according to World Bank data, the current per capita emissions rate in the United States is 16.5 tons of carbon dioxide per person per year.[56] You can do the math.

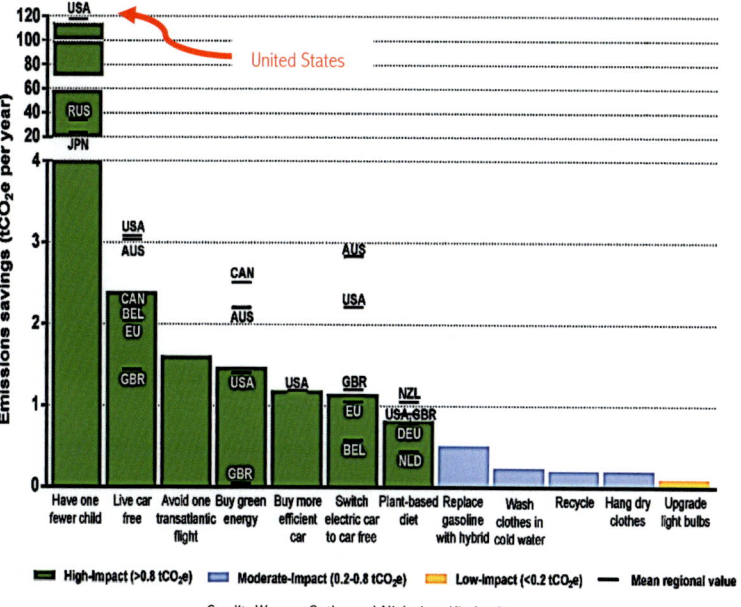

EMISSIONS SAVINGS: MOST EFFECTIVE INDIVIDUAL ACTIONS

Credit: Wynes, Seth and Nicholas, Kimberly
https://iopscience.iop.org /article/10.10881748-9326/aa7541

GEOGRAPHIC DISTRIBUTION OF POPULATION GROWTH: In most of the sustainability-population models, the primary demographic variable considered was population size. But our per capita impact on the environment is not equal around the globe, as the spectrum of consumption and production is very wide—as a result of broad disparities in economic, social, and political conditions. The location on the planet where population growth rate reductions take place is a critical sustainability factor. A simple example: We empirically know that the least developed and poorest nations produce the fewest harmful emissions per capita, while the industrialized, most affluent nations produce the greatest emissions per capita. The United States, with 4.25% of the world's population, produces approximately 14% of the world's carbon dioxide emissions; while India, with 17.7% of the world's population, produces only 7% of the world's carbon dioxide emissions.[57] Consequently, a population reduction in India would have a very different immediate impact on greenhouse gas emissions (and all of the other environmental impacts we are exploring) than an equal population reduction in the United States. As a result, we can obtain much greater environmental benefit from fertility reductions in our most industrialized, highest-consuming, and heaviest-polluting countries.

GRAY POPULATION BOOM: Due to advances in medical technology, improved nutrition, access to education, and other lifestyle enhancements we enjoy in the twenty-first century, life expectancy has drastically increased around the globe. According to the UN-Population Division, average global life expectancy in 1800 was 28.5 years, while in 1950, it reached an average of 45 years, and by 2019 average global life expectancy was 72.6 years. In the most developed and affluent countries like Germany, it is even higher than that average—closer to 81 years, and in the least developed countries, life expectancy is substantially lower than the average—closer to 55 to 60 years.[58] At either extreme, life expectancy has been increasing at a very fast pace all over our globe.

As a result of (i) global increase in average life expectancy, (ii) very high fertility and birth rates during the decades following World War II, and (iii) present-day reductions in fertility rates in most areas of the world, almost all countries today have a disproportionate number of elderly people in their populations. The number of people aged 65 or older is projected to *double* by 2050, while the number of people aged 80 or over will *triple* over the next 30 years.[59]

There is research that concludes that the growing elderly segment of the population will increase harmful emissions, and there is equally credible research that concludes just the opposite. In either case, recent natural disasters reflect that elderly populations are more vulnerable to the effects of climate change. The majority of deaths from Japan's earthquake in 2011 and Hurricane Katrina in 2005 occurred among older people. There was a heatwave in Europe in 2003, after which BBC news reported 14,800 deaths in France, of which 70% were people aged over 75. In addition, public health research suggests that environmental emissions and toxins disproportionately compromise the health of the elderly.

YOUTH EXPLOSION: Life expectancy's first cousins are infant and child mortality and fertility rates. *Infant and child mortality* is generally defined as the number of deaths before reaching the ages of one and five, respectively, per 1,000 infants or children. Today, it is difficult to process the fact that in 1800, even in the most advanced and affluent countries, 33% of the children born died before attaining five years of age. By comparison, the World Health Organization reports that in 2018, the global average rate of children dying before age five was about 4%. Still, the average rate of child mortality in developing countries is approximately 14 times greater than the rate of more-developed countries (e.g.; 7.4 per thousand in Afghanistan in 2015 compared to .43 per thousand in Australia/New Zealand in 2015)[60]

Fertility is defined as the average number of births women in a defined geographic area will have during their childbearing years. Fertility tends to be inversely related to the socioeconomic development of the female population, but in particular the autonomy, status, education, and well-being of women. Western and Northern Europe and North America reached a replacement rate of fertility (2.1 children per woman) at a time when fertility had only started to decline in parts of Asia, India, Africa, and Latin America. On the other hand, the average fertility rate of sub-Saharan African countries remains at 4.6.[61] As a result of the very high fertility rates, certain areas, primarily in India, sub-Saharan Africa, and South Asia, are experiencing a youth boom. A disproportionate share of their populations is under age 15, and their respective birth rates are much higher than the global average. The map and graphic on the following page from Visual Capitalist.com shows the median age for each continent, with the African continent having the youngest population. A very high rate of population growth in these countries will continue until the youths and their offspring born during this boom are no longer of childbearing age. Some of these very same areas have very low per capita availability of fresh water and arable land, making it even more difficult to feed their growing populations. To compound that challenge, with population growth concentrated in the less-developed regions, the number of people and countries lacking the resources to adapt to climate change will continue to increase. These countries lack not just the financial resources, but also the public institutions, infrastructure, and technology to implement adaptive measures.

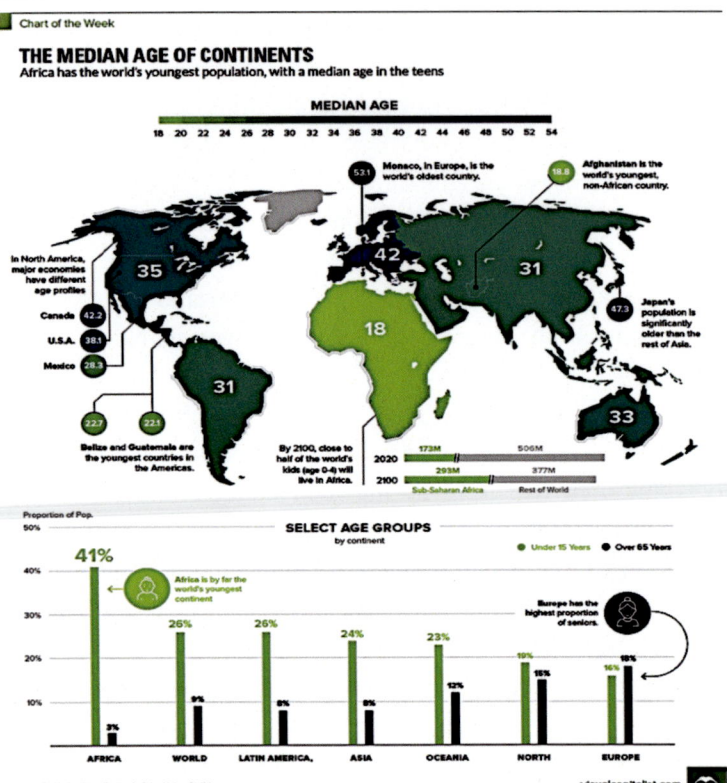

THE MEDIAN AGE OF CONTINENTS
Africa has the world's youngest population, with a median age in the teens

Image from Visual Capitalist, an open-source website. Credit: Jeff Desjardins 2019.
https://www.visualcapitalist.com/mapped-the-median-age-of-every-continent/

URBAN MIGRATION: Another critical demographic trend is the worldwide movement to cities. In 2016, 55% of the world's population was living in urban areas. By the middle of this century, 67% of the world's population is expected to be living in urban areas, and by the end of this century, almost the entire world's population will be living in cities. In 1803, when our planet experienced its first billion in population, only 3% of the population was living in cities. Urban areas can offer superior economic opportunities, quality education, and access to excellent health care. As you might guess, the beginning of the global movement from rural areas to the cities began during the Industrial Revolution.

Today, this worldwide urban shift is creating megacities, defined as cities with populations in excess of 10 million. We easily recall New York and Tokyo as the first recognized megacities, having reached the 10 million threshold in the 1950s. Today, there are at least 28 megacities, most of which are located in Asia and Africa. It is estimated that by the year 2100, 13 of the world's 20 largest megacities will be located in Africa, and three will be located in India. Cities like New York and Los Angles are not projected to show up as one of the world's top 20 largest cities.

Sustainable urbanization on this scale is going to be a challenge for even the most advanced and affluent areas of the world. This is a burden that will fall in large part to governments, but also to public-private partnerships. Cities are hotbeds of economic activity, and their densely packed populations consume food and energy and produce waste in solid, liquid, and gaseous forms. There is frequently not sufficient space to produce food and raw materials at the local level, requiring that foods, raw materials, and waste products be transported in and out of the city. As a result of large amounts of heat-absorbing asphalt typically found in cities, such areas tend to be warmer than adjoining rural areas. Related to this is the preponderance of impervious surfaces and building materials in our cities that do not permit the cooling effect from evaporating water. Finally, such impervious surfaces do not permit rainwater to be reabsorbed into the earth; instead, they give rise to large quantities of stormwater runoff.

We examined Ehrlich's formula for environmental impact earlier, which reads as follows: $I = P \times A \times T$. Based on this formula, one would expect a megacity to have geometric increases in adverse environmental impacts as a city grows in both population and economic output. But, it turns out that environmentalism and urbanization are not necessarily mutually exclusive. Dense urban areas can have a smaller ecological footprint than diffuse suburban areas for a given population size. There can be energy savings derived from dense and multifamily housing, compared with sprawling suburbs of single-family homes. Cities tend to be more walkable and generally have user-friendly mass transportation options that reduce dependence on automobiles. And finally, densely populated urban areas can mean that there is more open space available for wildlife habitat, farmland, forests, and conservation areas outside the city. Interestingly, in his book *Green Metropolis*, David Owen lauds the dense, concentrated canyons of Manhattan. He considers Manhattan as one of the greenest places in America, where most people live in modest-sized apartments and walk or use mass transit to shop, go to work, and recreate.[62] It appears that as some urban regions increase their populations, their per capita emissions actually decline. Research performed by Dr. Marilyn Brown for the Brookings Institute concluded just that.

> When using population to measure metropolitan size, the coefficient of 0.92 indicates that the effect on total carbon emissions of increasing population size is slightly less than proportional: more precisely, a 1% increase in population is associated with a 0.92% increase in carbon emissions. When using economic output to measure metropolitan size, the scaling coefficient is 0.79, which means that a 1% increase in economic output is associated with only a 0.79% increase in CO_2 emissions. Together these two results tell us that larger metropolitan areas are indeed more energy efficient than smaller metropolitan areas—and in particular that increasing metropolitan wealth (as measured by GMP) is associated with decreasing energy consumption. The energy metabolism of metropolitan areas slows down with increasing size.[63]

It is logical for cities to take the lead in planning for buildings with minimal environmental impacts, as you can optimize the per capita environmental impact. Interestingly, this same research from the Brookings Institute found that the 40 largest megacities produce 66% of total global economic output and 90% of global innovation, while housing just 18% of the world's population.[64] That is serious food for thought: 18% of the world's population producing 90% of global innovation and 66% total economic output.

So, in the final analysis, environmental sustainability and urbanization may not be mutually exclusive. This is exceedingly good news, as most of us are or will be living in urban areas by the end of this century. Of course, the environmental drawbacks of megacities can be as substantial as the cities themselves. By way of example, in 1960, Lagos, Nigeria was a small, coastal city that was well-governed and orderly. Between 1960 and today, Lagos grew from fewer than 200,000 people to nearly 20 million. Today it is one of the world's largest megacities, sprawling over nearly 450 square miles. It has several vastly wealthy areas and numerous millionaires; however, much more of Lagos is impoverished. More than half the Lagos residents (as much as 66%) live in informal settlements or slums. The great majority of Lagos residents are not connected to piped potable water, electricity, or sewers. The city's streets are choked with traffic, its air is full of fumes, and it produces more than 10,000 metric tons of waste per day. The history of Lagos is a cautionary tale reminding us that when megacities grow at rates that do not allow for infrastructure to keep pace with growth, the result is widespread slums, sickness, and crime.

WEALTH AND INCOME: People around the world consume resources differently and unevenly. An average middle-class American consumes 3.3 times the subsistence level of food and almost 250 times the subsistence level of clean water. So, if everyone on Earth lived like a middle-class American, then the planet would only have a carrying capacity of around 2 billion people and would require nearly four Earths to support the current population. History has shown us that as soon as developing countries begin to grow their economies and develop an expanding middle class, those citizens begin to consume like middle-class Americans. Countries have been observed to pass through defined stages of economic development from subsistence level agriculture to mass industrialization to information technology. The less-developed countries of our planet will, in time, experience economic growth, with the only question being at what rate. Of course, it is desirable for less-developed countries to move through these stages of economic development, but how should the more-developed world address the increased production and consumption that will accompany this economic growth?

Many of the less-developed countries that are least prepared to deal with population growth and environmental mitigation are former colonies of today's developed countries. A share of the wealth of developed countries can be attributed to past exploitation of cheap labor and resources found in these former colonies. I clearly remember many of these less-developed countries gaining their independence when I was in grade school, which is not really *that* long ago. Consequently, these former colonies only began to function autonomously in the 1950s and 1960s, while the colonial powers had hundreds of years head start in developing their economies and infrastructures. What is the nature of our obligation to the people of these countries?

RELIGIOUS AND CULTURAL VALUES: Population dynamics are strongly affected by religious and cultural values that favor and encourage reproduction. The best examples include religious or state bans against all or some birth control methods. Most religions endorse and encourage reproduction. Technical advances in medicine have made available a wide range of safe and effective contraception choices for family planning; still, unplanned and unintended pregnancies continue to occur in large numbers, particularly in less-developed countries. Exacerbating the challenges of family planning in some less-developed areas are early marriage expectations, women's lack of autonomy in reproductive matters, and a lack of male involvement in matters of sexual and reproductive health. Some more-developed countries are experiencing shrinking populations due to low fertility rates, high emigration rates, and a large share of post-childbearing adults—these include Japan, Australia, Canada, Denmark, Singapore, Russia, Poland, Czech Republic, Hungary, and Germany. Responding to the loss of producers and consumers leaders believe are necessary to keep their economies robust, some of these countries are offering women and families financial incentives for bearing children. The incentives come in the form of tax rebates, paid maternity leave, child care assistance, housing assistance, as well as direct cash payments on a per-child basis.

EDUCATION OF WOMEN: The graphic on the following page dramatically depicts the compelling inverse relationship between women's education and fertility rate in selected countries. The red bars reflect educational attainment, and the green bars reflect the number of children. Educated women tend to marry later and have fewer and healthier children. Is this a causal relationship or merely a correlation? Elina Pradhan, of the Harvard T.H. Chan School of Public Health, has performed studies in sub-Saharan African countries to determine whether education does impact the number of children a woman will bear. The following is a summary of her research.

We investigate the effect of female schooling on teenage fertility using an education reform in Ethiopia in 1994 as a natural experiment that led to a jump in female school enrollment and about 0.74 years of additional schooling for the first two exposed cohorts. Using a regression discontinuity approach, we find that each additional year of schooling lowers the probability of both teenage marriage and teenage childbearing by about six percentage points. This casual estimate is consistent with the steep gradient of teenage marriage and fertility with education observed in the data.[65]

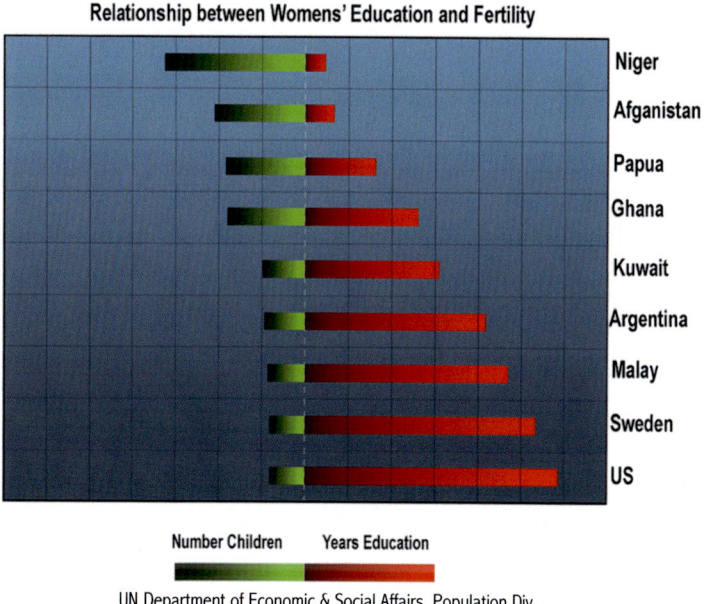

UN Department of Economic & Social Affairs, Population Div.
Estimates & Projections of Family Planning Indicators (2019)

The best way to reduce fertility rates in developing countries is to (i) ensure that all women have access to education, family planning information, reproductive health services, digital technology, training, and jobs, (ii) enforce laws that prohibit childhood marriage, (iii) outlaw practices that discriminate against women and girls, and (iv) enforce laws that protect women and girls from violence. The subject of overpopulation in less-developed countries is fundamentally about the rights of women to have autonomy over their bodies and life choices. The United Nations has long been a strong advocate for women's rights, and it has recently renewed its commitment to women in developing countries. António Guterres, the Secretary-General of the United Nations, recently stated the following in a speech at The New College in New York City.

> Dear students, dear friends,
>
> Gender equality is a question of power; power that has been jealously guarded by men for millennia. It is about an abuse of power that is damaging our communities, our economies, our environment, our relationships and our health. We must urgently transform and redistribute power, if we are to safeguard our future and our planet. That is why all men should support women's rights and gender equality. And why I am a proud feminist. Women have equaled and outperformed men in almost every sphere. It is time to stop trying to change women, and start changing the systems that prevent them from achieving their potential. Our power structures have evolved gradually over thousands of years. One further evolution is long overdue. The 21st century must be the century of women's equality. Let us all play our part in making it so.[66]

The Bill & Melinda Gates Foundation has committed millions of dollars to the goal of increasing education, empowerment, and health of women and girls in developing countries. They are channeling these funds to institutions that can participate in improving the reproductive health and choices for women in developing countries, where unplanned pregnancies, unsafe childbearing, and abortions are a significant cause of death and poor health. The Gates Foundation is working with the Johns Hopkins Bloomberg School of Public Health to build the infrastructure to achieve these goals. It is estimated that 120 million sexually-active women who want and need access to family planning services do not have it. Every year, there are approximately 66 million unwanted pregnancies and 20 million unsafe abortions, most of which occur in developing countries. So far, the lion's share of this effort remains in the hands of nongovernmental organizations, but it is well past the time for world governments to endorse this part of the solution and to take action.

BOTTOM LINE: POPULATION: We are in the throes of a population growth momentum that will take multiple generations to abate. Based on projected fertility rates, populations in the less-developed countries of Africa and South Asia are expected to increase significantly, aggregating in vast megacities, while populations in Europe and North America are expected to decrease. By 2050, more than 70% of the world's population is expected to live in Asia and Africa, while 10% of the world's population is expected to live in Europe and North America. Much of the Asian and African population will be located in less-developed areas that offer the least amount of opportunity for health and education services and job opportunities.[67] According to the World Bank, more than 40% of our global population, more than 3 billion people, currently lives on less than $5.50 per day. Our global population is not expected to stop increasing until 2100, when today's young females, their children, and grandchildren will have moved out of their childbearing years. By the time this occurs, we will have reached a projected global population of 11 to 16 billion. The wide range of this United Nations population projection is almost entirely a function of future fertility rates in these less-developed areas.

Gains in agricultural technology and improved efficiency in producing energy have extended the Malthusian limit for our planet; yet, we must not forget that many essential resources remain finite. While global population growth rate has been decreasing, we remain in the midst of a population momentum blizzard that cannot be stopped or turned around abruptly. While our technology allows us to support more and more people with each of our finite resources, common sense tells us that there is an actual limit to how long technology can keep the Malthusian limit at bay.

Before we pat ourselves on our collective backs for our decreasing population growth rate, we must face a really uncomfortable reality. I am just going to say it because you have probably reached the same conclusion by now: Our planet is overpopulated--we have already overshot Earth's carrying capacity with our current population. It is only because so many millions live in poverty in less-developed regions that those of us privileged to live in the more-developed and affluent regions of the world are not having to defend our comfortable access to the planet's diminishing resources. If we all lived like the average European, then Earth could support about 3 billion of us.

As mentioned earlier, the former colonial powers of the developed world had a significant head start in developing their economies and infrastructure, compared to the countries that made up their former colonial holdings. As with atmospheric carbon dioxide, where merely reducing emissions is not enough to stop planetary warming, the only path to stopping or slowing population momentum is to reduce the average fertility rate below the replacement rate. I hope that it is obvious that this must be done on a voluntary basis through education and communication. The lowest hanging fruit lies in the developed world, where each ecological footprint equates to many footprints in the less-developed world. For instance, according to the Global Footprint Network, the United States' per capita footprint, expressed in global hectares per person, is 8.22, while India's per capita footprint is 1.16. United Nations-Population Division currently projects North American and European fertility rates to remain flat until the end of the century, at about 1.8 to 1.9. If the North American and European populations could achieve fertility rates of 1, our planet would emit a huge sigh of relief as our aggregate world ecological footprint and population growth momentum greatly improve.

We will leave our discussion of population with a quote from one of my favorite scientists, Harvard University sociobiologist Edward O. Wilson, followed by a comment on Wilson's quote by journalist Natalie Wolchover:

> The constraints of the biosphere are fixed. The bottleneck through which we are passing is real. If everyone agreed to become vegetarian, leaving little or nothing for livestock, the present 1.4 billion hectares of arable land (3.5 billion acres) would support about 10 billion people.

> The 3.5 billion acres would produce approximately two billion tons of grains annually. That's enough to feed 10 billion vegetarians, but would only feed 2.5 billion US omnivores, because so much vegetation is dedicated to livestock and poultry in the United States.[68]

Chapter 5

FOOD PRODUCTION IMPACTS

LIVESTOCK AGRICULTURE

I hope that by now I have made the case that because of population growth momentum, (a) we have an ever-increasing population, notwithstanding global reductions in average fertility rates and birth rates, and (b) without negative population growth (in which fewer humans are joining us than leaving us), we will continue on a trajectory that may take longer to correct than we have time and resources. The United Nations Food and Agriculture Organization (FAO) has stated:

> By 2050 the world's population will reach 9.1 billion, 34 percent higher than today. Nearly all of this population increase will occur in developing countries. Urbanization will continue at an accelerated pace, and about 70 percent of the world's population will be urban (compared to 49 percent today). Income levels will be many multiples of what they are now. To feed this larger, more urban and richer population, food production (net of food used for biofuels) must increase by 70 percent. Annual cereal production will need to rise to about three billion tonnes from 2.1 billion today, and annual

meat production will need to rise by over 200 million tonnes to reach 470 million tonnes.[69]

Our planet has neither sufficient arable land nor water to accomplish that objective if our per capita consumption of animals continues at the same rate as today. Consider the following highlighted realities of livestock agriculture:

- *Water:* In most areas of the planet, 70% of total global fresh water withdrawals are used for agriculture. One-third of that water is used to grow grains to feed the world's 70 billion livestock animals we slaughter each year. Meanwhile, 29% of our fellow humans do not have convenient access to safe drinking water.[70] It is estimated that 50% to 64% of the world's population currently lives in water-stressed areas where they experience severe water scarcity at least one month of the ear.[71]

- *Food:* Of all the grain products grown in the world, 45% is fed to the 70 billion livestock animals we slaughter each year worldwide. Meanwhile, in 2013, 840 million people, or almost 12% of our global population, were undernourished (did not receive adequate calories), and 2 billion people were deficient in essential nutrients derived from food.[72]

- *Forests:* Tragically, 70% of the Amazon rainforests (a significant carbon sink and source of oxygen) have been slashed and burned to create open land to graze livestock and grow grains to feed livestock, while the humans working those fields frequently go hungry.[73] We have globally destroyed approximately 47 million acres or 74,000 square miles per year of forested land since 2000. This is almost the size of the state of Nebraska in the United States.[74]

- *Land*: Livestock agriculture is using 77% of all arable land around the globe, either for grazing livestock or for growing crops for livestock. This amounts to a significant portion of the total ice-free land surface of the planet.[75]

- *Waste:* The waste from 2,500 dairy cows on a single farm is equivalent to the waste from a city housing half a million people. Most of this waste goes untreated, and our technology for keeping this waste from polluting water sources and our atmosphere is severely deficient.[76]

- *Greenhouse Gases:* According to the FAO, livestock agriculture contributes 18% (by volume) of our annual greenhouse gas emissions, which is more than the entire transportation sector—all trains, planes, buses, cars, and ships. Other respectable estimates are much higher (see page 112). Still, we spend much of our collective time and resources addressing carbon dioxide emissions from our cars. Go figure . . . Livestock agriculture emissions account for 37% of total methane emissions, 65% of total nitrous oxide emissions, and 64% of total gaseous ammonia emissions.[77]

- *Soil:* We are eroding our farmland soil 10 to 40 times faster than it can naturally regenerate. Stated another way, we lose about 24 million acres of cultivated farmland every year due to soil erosion (and urban sprawl).[78]

As you can see, livestock agriculture has a profoundly egregious impact on almost every aspect of our environment. According to estimates by the United States Department of Agriculture (USDA), the average American ate 222 pounds of meat in 2018—we are literally eating our weight in meat. There is no question that the demand for meat and dairy products is increasing at warp speed. The demand is being driven by rising incomes, growing populations, subsidized meat prices, and urbanization. At the same time, the policy response to reduce the adverse environmental impacts of livestock agriculture lies somewhere between slow and nonexistent. In industrialized countries, livestock agriculture is big business, buffered by long-term subsidized relationships between governments and livestock conglomerates. Currently, this sector enjoys numerous price supports that subsidize this industry, many of which were developed almost 100 years ago to provide relief to small farmers during times of drought or economic hardship. Taxpayers in the United States are currently contributing $38 billion to subsidize meat, dairy, and eggs. The countries of the

European Union have similar supports. Speaking of government protection of an industry, do you remember when the members of the Texas beef industry sued Oprah Winfrey for publicly announcing in 1996 that learning about England's mad cow disease (BSE) "stopped me cold from eating another burger." At one point, 13 states in the United States had similar "meat disparagement" laws that made it illegal to make critical comments in public about meat products.

The United Nations has attempted to take a leadership role in moving livestock agriculture toward more sustainable practices. As far back as 2006, the United Nations, through its FAO and its Livestock, Environment, and Development Initiative, published *Livestock's Long Shadow*, a book outlining the challenges posed by livestock agriculture. The executive summary begins with the following:

> The livestock sector emerges as one of the top two or three most significant contributors to the most serious environmental problems, at every scale from local to global. The findings of this report suggest that it should be a major policy focus when dealing with problems of land degradation, climate change, and air pollution, water shortage and water pollution, and loss of biodiversity. Livestock's contribution to environmental problems is on a massive scale, and its potential contribution to their solution is equally large. The impact is so significant that it needs to be addressed with urgency. Major reductions in impact could be achieved at reasonable cost.[79]

LIVESTOCK GREENHOUSE GASES: One of the largest producers of methane is animal agriculture, animal husbandry, or call it what it is: breeding and raising animals so they can be slaughtered for humans to eat. Currently, there are approximately 70 *billion* farm animals slaughtered worldwide each year to produce meat products for humans to eat, two-thirds of which spend the last months of their lives in crowded and filthy factory farms. Livestock, particularly ruminants like cows and sheep, emit vast quantities of methane and nitrous oxide as food passes through the various compartments of their stomachs. This process is called enteric fermentation, a process by which microbes decompose and ferment food in the digestive tract or rumen, producing methane as a byproduct. When we look at greenhouse gas comparisons, the charts rarely take into account the relative impact or GWP of each of the various greenhouse gases. We learned earlier that methane and nitrous oxide molecules in the atmosphere are far more potent than carbon dioxide because they trap larger quantities of heat.

The United Nations FAO has estimated that 18% of all global greenhouse gas emissions are from the agriculture sector, and the IPCC has estimated this amount to be 24%; however, both these estimates were credibly challenged by Robert Goodland and Jeff Anhang, of the Worldwatch Institute. Goodland and Anhang make the argument that the actual greenhouse gas emissions from livestock agriculture are about 51% of total emissions, if you take into account the respiration of the billions of livestock animals, the loss of carbon sinks by deforestation, the relative GWP of methane, and impacts of fish aquaculture.[80] California apparently received the memo about livestock emissions, as it passed a law in September of 2019 requiring California farmers to reduce greenhouse emissions from livestock by 40% (using 2013 levels as a baseline) over the next ten years.

DEFORESTATION: Next on the list of livestock agriculture offenses is the slashing and burning of forestlands to clear for agriculture; that is, to grow grains for animals to eat so we can eat the animals. The conversion of forests into agricultural land is one of the primary causes of deforestation, and it is driven by demand for livestock feed crops, grazing land for livestock, and palm oil for processed foods. We explored the carbon cycle earlier, whereby plants and trees take up free carbon dioxide from the atmosphere, separate the carbon from the oxygen during photosynthesis, store the carbon, and give off molecular oxygen for us to breathe. The carbon is "fixed" in the soil and leaves, trunk, and branches of the tree or plant. Consequently, when we burn acres of forestland to replace with agriculture crops, we get a triple whammy for our efforts: (i) we reduce the amount of critical biomass available to absorb atmospheric carbon dioxide, (ii) we reduce the amount of biomass available to produce oxygen for us to breathe and (iii) we detonate what some have referred to as a "carbon bomb." The carbon bomb is a product of decades of stored carbon released into the atmosphere as the trees burn, and produce more greenhouse gases than any belching coal plant you have seen.

The Amazon rainforest is the largest contiguous rainforest in the world, making it Earth's most significant carbon dioxide sink. Large areas of Amazonian rainforest are cleared every year, using the slash-and-burn technique. The Brazilian National Institute for Space Research observed via satellite more than 75,000 wildfires during the first eight months of 2019. It is estimated that at least 70% of all deforestation of the Amazon rainforest is for purposes of growing crops for animal feed. The World Resources Institute estimates that tropical tree cover can provide as much as 23% of the carbon mitigation needed to meet the voluntary goals agreed to by world leaders in 2015 in Paris. The Institute also has stated that if tropical deforestation were a country, it would rank third in carbon-dioxide-equivalent emissions, behind China and the United States.[81] The World Wildlife Organization estimates that more than 80% of Earth's animal species and 66% of the world's plant species live in forests, and the planet has already lost many species due to the extinction of animals from lost habitat.[82] We have not yet discovered all of the animal and plant species of the rainforests, yet we are losing hundreds of species every day. Ethnopharmacologists are searching the rainforests of the world for plants that can be used to produce tomorrow's anticancer drugs, but where will these plants be found when we have destroyed most of our tropical rainforests?

LIVESTOCK WATER CONSUMPTION: One of the most significant environmental impacts of agriculture, but animal agriculture, in particular, is excessive water use. There is no way around the fact that raising livestock is a water-intensive business, whether you are a small organic farmer grass-feeding your livestock or a corporate conglomerate owning multiple large concentrated animal feeding operations (CAFOs). Agriculture uses 70% of US fresh water supplies, but more than 30% of the US agricultural water use is directed to growing food crops for livestock so that humans can eat the livestock. Water is needed, not only to grow crops for the 70 billion slaughtered livestock animals worldwide, but also to hydrate those animals, to wash down their pens, and to manage their manure. In the United States, animal agriculture consumes 36 to 74 *trillion* gallons of water annually. By comparison, much-disparaged hydraulic fracturing (fracking) uses "only" 70 to 140 *billion* gallons annually. I mentioned earlier that the United Nations Water agency (UN-Water) has estimated that 50% to 64% of the world's population currently lives in water-stressed areas. We direct much of our water conservation efforts to doing everything possible to conserve water at home by installing low-flow showerheads and faucets and 1.6-gallon toilets, taking two-minute showers, and restricting the use of dishwashers and washing machines. I have to ask: If it takes 660 gallons of water to produce a single quarter-pound hamburger, and we can only save about 47 gallons of water per *month* by implementing all of the conservation efforts just described, what are we thinking? A typical month's worth of hamburgers (e.g., 4 quarter pounders) consumes 4.5 *years* of water savings. Stop and think about that for a minute.

Food Water Usage
from the Water Footprint Network.org, per terms of use

Food Item	Serving Size	Water Footprint
Steak (beef)	6 ounces	674 gal
Hamburger	1	660 gal
Ham (pork)	3 ounces	135 gal
Eggs	1 egg	52 gal
Salad	1	21 gal
Coffee	1 cup	34 gal
Wine	1 glass	34 gal

LIVESTOCK MANURE: Livestock manure management is a critical environmental issue relating to water and air quality. "Over 1 *billion* tons [emphasis added] of animal manure is produced annually in the US. Animal manure is an excellent plant nutrient source and soil amendment when properly treated and applied. Manure contains plant macro- and micronutrients, supplies organic matter, improves soil quality, and maintains or increases soil pH in acid soils. Following a nutrient management plan and proven best management practices will improve manure nutrient use efficiency and reduce the impact of the land application of manure on water quality."[83] But that is not exactly what is happening in the United States and many other countries.

On the other hand, some countries are very advanced in regulating and using technology to manage livestock manure. I came across hundreds of textbooks and research papers written on the subject, suggesting that manure management might be a vital issue. The Wageningen University and Research Institute in the Netherlands is considered one of the best universities for agricultural sciences. Much of their research is open source, so their website is a cornucopia of information about improved agricultural practices. The website can be found at https://www.wur.nl/en/About-Wageningen.htm.

In the United States, most cattle spend the last portion of their lives in densely populated indoor facilities or outdoor feedlots, where they can be fattened as quickly as possible for market. Poultry, pigs, and cattle together produce 335 *million tons* of waste per year in the United States. While this waste is potentially rich organic matter that can help crops grow, there is typically much more manure produced than can be repurposed. Some of these indoor facilities have the cattle standing on grated floors, which allow the manure to drop to the holding tank below the floor. The manure collected in the tank below is then piped or otherwise moved to a large manure lagoon, a concrete basin that allows the manure to decompose. As the manure decomposes anaerobically (without oxygen), it produces methane, ammonia, toxic hydrogen sulfide, nitrous oxide, and carbon dioxide. Anaerobic manure lagoons raise numerous environmental concerns, such as groundwater pollution when manure seeps out of the lagoons and greenhouse gas emissions of methane and nitrous oxide. On the other hand, aerobic lagoons produce less odor and create an end product that can be used as a rich fertilizer for crops. Aerobic lagoons require more space and are more costly to build and operate. Aerobic lagoons require aeration of the manure, but the process transforms raw animal manure into a crop-ready liquid fertilizer that can be sprayed directly onto growing crops without odor, pathogens, or plant stress. You are probably wondering how much manure various types of livestock generate per day; if so, please see the table on the following page. In the United States, we are simply producing more manure than our current technology can manage.

Manure Production: Pounds per Day per 1000lbs

Livestock type	Total Manure	Nitrogen	Phosphorus
Beef[1]	59.1 lbs	0.31	0.11
Dairy[2]	80.0 lbs	0.45	0.07
Hogs and pigs[3]	63.1 lbs	0.42	0.16
Chickens (layers)	60.5 lbs	0.83	0.31
Chickens (broilers)	80.0 lbs	1.10	0.34
Turkeys	43.6 lbs	0.74	0.28

[1]High forage diet. [2]Lactating cow. [3]Grower.
Source: USDA Natural Resources Conservation Service. Agricultural Waste Management Handbook (1992)

LIVESTOCK LAND USE: Take another look at the land-use graphic in Chapter 2, which shows that we have available 9.5% of the planet as potentially arable land for food crops. As mentioned earlier, 77% of that arable land is used to raise livestock food crops and for livestock grazing, leaving us with 23% for the cultivation of food crops for humans. It also shows that the portion of arable land used for livestock food crops and grazing produces only 18% of the world's total calories. Looking at this another way, one acre of land can yield 250 pounds of dressed beef, as compared with 50,000 pounds of tomatoes, 53,000 pounds of potatoes, and 30,000 pounds of carrots.[84]In research from the University of Edinburgh that attempted to quantify the effects of animal agriculture on available land use, the researchers found that if the entire world ate a diet that was similar to the average diet in the United States, that we would need 178% *more* land to produce that food. On the other hand, they found that if the entire world ate a diet similar to the average diet in India, global demand for food could be satisfied with 55% *less* land. This analysis was based on the 2016 world population and did not attempt to quantify additional land required in the future to feed a growing population.[85] The referenced article is a fascinating read, and I urge you to take a look at it. The authors have developed a scale that allows them to measure the effect of various diets and dietary shifts on available agricultural lands.

LIVESTOCK WATER POLLUTION: Animal agriculture is the primary source of water pollution by nitrites, phosphates, insecticides, and pesticides. In particular, the contamination of groundwater by agricultural chemicals and waste products is a significant issue in both industrialized and developing countries. Feed crops for livestock are a significant contributor to this process. According to the United Nations FAO, livestock agriculture in the United States is responsible for 55% of the erosion and sedimentation, 37% of the pesticide use, 50% of the antibiotic use, and 33% of the nitrogen and phosphorus runoff into fresh water.[86] The global livestock business is among the planet's most destructive economic sectors, contributing, among other things, to water pollution from animal wastes, antibiotics and hormones, chemicals from tanneries, and fertilizers and pesticides from spraying feed crops.[87]

Livestock agriculture is also responsible for eutrophication in coastal and freshwater ecosystems. In the section on nitrogen, we explored how plant and animal life of freshwater ecosystems and our oceans are suffocated to death by algae blooms. When agricultural nutrients flow into local waters, they provide super nourishment for normally existing algae. These normally existing algae feed on the agricultural nutrients, then they multiply like crazy and compete with existing plants and animals for oxygen. The algae usually win.

LIVESTOCK ANTIBIOTIC, HERBICIDE, AND PESTICIDE USE: Worldwide, more than 160 million animals are slaughtered for human consumption every day. While consuming those animals, we are also consuming that animal's portion of the 13.6 million kilograms per year of antibiotics given to livestock in the United States. These antibiotics are fed to the animals to promote growth and to prevent the spread of diseases arising from the unhygienic conditions of crowded factory farms. According to the FDA, this amount is nearly four times the quantity of antibiotics sold annually for human consumption. Then we wonder why scientists are encountering antibiotic-resistant bacterial infections. It would be an unfortunate irony if one of those antibiotic-resistant bugs ends up being the great Malthusian leveling agent of our overpopulated planet.

MILK PRODUCTION: The dairy industry would like us to picture its cows, perhaps a few goats, happily grazing in the sunshine on a beautiful green landscape, with a blushing milkmaid in the foreground. This is a happy image, but it is not exactly the image of the current industrialized dairy industry. Like all mammals, cows must be pregnant or recently have delivered a calf before they can produce milk. Dairy cows are forcefully impregnated, over and over, from a very young age and forced to remain pregnant as long as they are able. They are impregnated by semen mechanically withdrawn from a bull and then forced into the cow while she is restrained in a squeeze chute. The cows are fed reproductive hormones and antibiotics to ensure that they will produce as much milk as possible. Normally, a pregnant cow would carry about a half a gallon of milk (3.5 pounds) in her udder, but farmers can get their cows to produce as much as 5 gallons of milk (35 pounds). The strain on the udder and the weight are significant, making for a very shorted life span. The calves are taken from their mothers as soon as they are born, much to the obvious distress of both mom and calf. Many of the males from those forced pregnancies are slaughtered shortly after birth since that is less costly than raising them to adulthood. Alternatively, the males are raised in confined cages for about four months and then sent to slaughter to become highly-prized veal.

The description above does not address the environmental impacts of the dairy industry, but I included it because so many of us, vegetarians included, consume dairy products with an illusory Madison Avenue advertising image in mind. We have discussed land use, manure management, methane emissions, water use, and water pollution associated with breeding and raising animals for slaughter for consumption by humans. Every one of those environmental impacts applies equally to the breeding and raising of cows for milk. The dairy industry has become industrialized, just like the meat production industry. Dairy cows spend most of their lives in an indoor stall, eating genetically modified grains instead of grasses. They are connected to mechanized milking machines. Much of the genome of the dairy cow has been engineered to ensure a high fertility rate, large, well-positioned udders, and strong milk production. Like beef cattle, dairy cows are fed antibiotics and hormones to promote growth and stave off infections from the unsanitary environment. Almost 90% of our milk is produced on 26% of the large dairy farms. This industry consolidation amplifies the environmental hazards associated with manure management, the need for large quantities of water, methane emissions, and water pollution. In response to public pressure, many dairies are offering milk from grass-fed cows raised without the use of hormones or antibiotics from small, local farms. Still, profitability requires even these farms to separate cows from their calves at birth, and many of the other practices described above.

FARM SUBSIDIES: The United States government provides $38.4 *billion* in subsidies to the meat and dairy industries, but only $17 *million* is provided to the fruit and vegetable industries. [88] The enormous influence of meat industry lobbyists has permitted these hundred-year-old subsidies to continue. The subsidies allow the recipients in the meat and dairy industries to sell meat and dairy products for less than the cost to produce. This, in turn, encourages us to buy more meat and dairy than we might otherwise buy if the products were sold at their actual retail price. Such subsidies can also lead to the following results:

- Small, nonsubsidized farmers are put out of business, as they cannot compete with the below-cost subsidized factory farm prices.

- Smaller local farms are replaced by large corporate farms, with a loss of the income to the local community when supplies are purchased elsewhere, and fewer employees are required.

- US corporate meat producers dump low-priced product on global markets, making it more cost-effective for global markets to import American products than produce their own. This, in turn, eliminates local small-scale farming opportunities in those foreign markets.

- The health care costs associated with consumption of excessive meat and dairy has produced a health crisis in developed countries, as conditions such as obesity, diabetes, cardiovascular disease, and cancers have become the norm. In the United States, these lifestyle diseases are responsible for two-thirds of all deaths and about $700 billion in direct and indirect economic costs each year.

- Finally, the increasing demand for meat and dairy is creating a growing industry that may be doing more harm to our planet than any other single source. The current prices for meat and dairy fail to take into consideration any of the costs of the previously mentioned factors. Instead, the costs are passed on to all of us, omnivore or vegetarian, in the way of increased health insurance premiums or taxes for environmental cleanup efforts.[89] Take a look at the protein consumption graphic on the following page. And let's not forget that it is *our* tax dollars that make up the subsidies paid to farmers.

LIVESTOCK AND PHOSPHORUS, AGAIN: We discussed earlier how phosphate strip mining is particularly destructive to the environment. Phosphorus is an essential nutrient for all plant and animal life. All living organisms require regular phosphorus intake. It cannot be replaced, and there is no synthetic substitute. Without phosphorus, there is no plant or animal life. Our dependence on phosphorous supplementation in agriculture began in the 1800s when farmers noticed that where phosphorus-rich guano (bird and bat excrement) was deposited in their fields, those areas experienced significant improvement in crop yield. The mining of phosphate began soon after that in the United States, as earlier described in Chapter 2. The availability of mined phosphate led to the current use of synthetic fertilizer. With the depletion of naturally occurring minerals in our topsoil by overuse and monoculture crops, agriculture could only produce about 60% of the food that it provides today if it did not use commercial fertilizers. We discussed earlier that estimates of existing reserves, based on median population growth estimates from the United Nations suggest that the planet has about 70 to 120 years of economically accessible phosphate reserves remaining. In light of these reserves and the lack of any substitute, does it make sense to use this precious mineral to grow food for livestock that we intend to slaughter and eat?

ANIMAL PROTEIN: It appears that *Homo sapiens* really enjoys animal protein and has done so since the earliest of times. Some researchers have concluded that eating meat played a critical role in our evolution, particularly in our brain development.[90] We did not even touch on poultry and swine production in this chapter, but both are burdened with similar environmental concerns. However, asking people to stop eating meat is not in the realm of possibility, and even the most logical, documented, and convincing presentation of the harm that animal agriculture is doing to our planet and our bodies is not going to stop *Homo sapiens* from eating meat. See the protein consumption graphic below showing how much more protein than needed by the body is being consumed by the developed world. There is nothing healthy about taking in more protein than required by the human body.

People Are Eating More Protein than They Need—Especially in Wealthy Regions

Image from World Resources Institute by Creative Commons

This is a true Malthusian dilemma, as meat-eating at current levels is not even remotely sustainable right now, and certainly will be even less so with projected increases in population and affluence in emerging economies.

FEED CONVERSION RATES: Not only are we eating far more animal protein than our bodies require for good health, the production of that protein before it reaches our plates, whether fish, fowl, or mammal, is hugely inefficient. Most of the caloric energy animals consume is used to fuel their metabolisms and to form bones, cartilage, feathers, fluids, and other body parts discarded when meat producers prepare meat for sale. The inefficiencies increase further when liquid weight is removed, including the weight of the water, blood, and other bodily fluids. Feed conversion ratios (FCR) are used to measure the quantity of eatable food produced per volume of animal feed consumed by the animal before slaughter. The ratio is calculated by dividing the weight of all of the feed administered over the lifetime of the animal by the weight of the fully dressed and prepared animal. For example, chickens are more efficient food converters than cows. They have a lower FCR than cows, meaning that it takes less feed to create a pound of chicken than a pound of beef. According to the FAO, demand for meat is expected to increase by 70% by 2030. Research led by Emily Cassidy in 2013 concluded that if we converted the arable land currently used for livestock (77% of total arable land) to grow food crops for humans, we could feed an additional 4 *billion* people. She adds that a full 36% of global calorie production is used as animal feed, but only 12% of those calories make it to the human body due to inefficient FCRs.[91] It is difficult to escape the conclusion that the impact of increasing global meat consumption is not sustainable at current levels and certainly is not sustainable with an increasing population and increasing demand from the existing population.

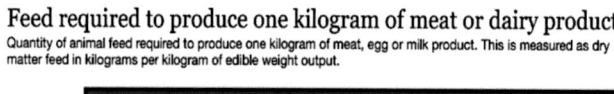

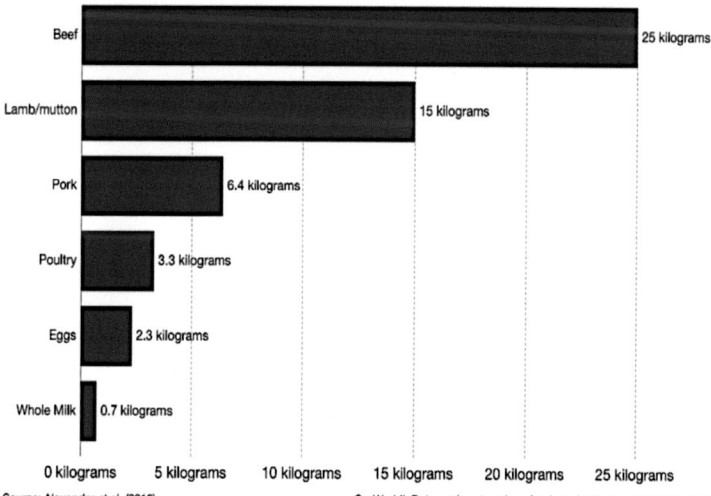

Source: Alexander et al. (2016)   OurWorldInData.org/meat-and-seafood-production-consumption/ · CC BY

Image licensed by Our World in Data, Creative Commons License by Attribution

ANIMAL WELFARE: The horrors of raising animals in factory farms have been documented over and over in film and in print. I mentioned some of these issues in our dairy farming discussion. I could not close this chapter without again mentioning the terror and torture these animals are forced to endure for our eating pleasure.

BOTTOM LINE: LIVESTOCK AGRICULTURE: Cattle and other livestock animals spend most of their lives grazing, but the final months of their lives are typically spent in concentrated feedlot facilities, where they are quickly fattened before they are slaughtered. It is here where most of the environmental hazards outlined in this chapter occur. So, the logical question that arises is why have we not returned to the traditional methods of raising livestock, grazing them until they are ready for market? A small percentage of our meat is currently produced this way, in an effort to eliminate some of the environmental and other horrors we have discussed. Unfortunately, such methods require more land, more time, and more water to prepare the animal for slaughter; thus, these methods can produce even more environmental harm. Some researchers argue that grazing cattle increases the health of the soil, making it able to sequester more carbon. In light of the additional land, time and water required, the need to quantify the possible sequestration benefit, and the cost of revamping an entire industry based on concentrated feedlot operations, the best short-term solution is simply to eat less meat and try to source the meat you do eat from ranchers producing meat more sustainably.

FISH AND AQUACULTURE

Yes, fish. Fish provide one of the best animal protein sources on the planet because fish are low in fat, have an excellent FCR, are loaded with omega-3 fatty acids, and offer a vast and diverse number of species. How many cow species can you name? We are currently taking more fish out of the ocean than can be naturally replenished by reproduction. The World Wildlife Fund (WWF) has stated, "The number of overfished stocks globally has tripled in half a century, and today fully 33% of the world's assessed fisheries are currently pushed beyond their biological limits." [92] Overfishing is closely tied to bycatch, the capture of unwanted sea life while fishing for a different species. "Bycatch is a serious marine threat that causes the needless loss of billions of fish, along with hundreds of thousands of sea turtles and cetaceans."[93] Also implicated in overfishing are government subsidies that offset the actual cost of doing business. These subsidies open the door for a large number of fishing vessels to continue fishing when it would not make economic sense to do so without this financial support. The WWF concludes that "Today's worldwide fishing fleet is estimated to be up to two-and-a-half times the capacity needed to catch what we actually need. The United Nations 2030 Agenda for Sustainable Development has called for an end to harmful subsidies."[94] Finally, "Some of the worst ocean impacts are caused by pervasive illegal fishing, which is estimated at up to 30% of catch or more for high-value species. Experts estimate illegal, unreported, and unregulated fishing provides criminals up to $36.4 billion each year"[95] Our wild fisheries must have time to regenerate.

Our global production of seafood was nearly 200 million tons in 2015 yet, according to The Ocean Foundation, we reached "peak fish" sometime in the late 1980s when we were producing about 90 million tons, or half as much as today. The oceans of our planet are not able to meet our demand for seafood, and our overfishing activities will undoubtedly harm marine ecosystems. As a result of the diminishing populations of fish in the oceans, finfish farming (also called aquaculture), is being developed at a brisk rate. According to the FAO, finfish farming is the fastest-growing food production sector in the world. Ironically, the growth of aquaculture is fueling more marine fishing to satisfy the increasing demand to produce fish meal and fish oil to feed to farmed fish. Sadly, 33% of all fish caught in the wild are used to make fishmeal to feed to livestock, aquarium fish, and farmed fish.[96] While aquaculture has the potential to feed our growing population, the rush to aquaculture is fraught with the risk that we will develop environmentally harmful systems.

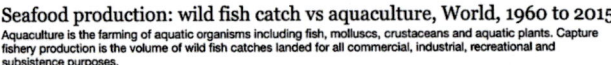

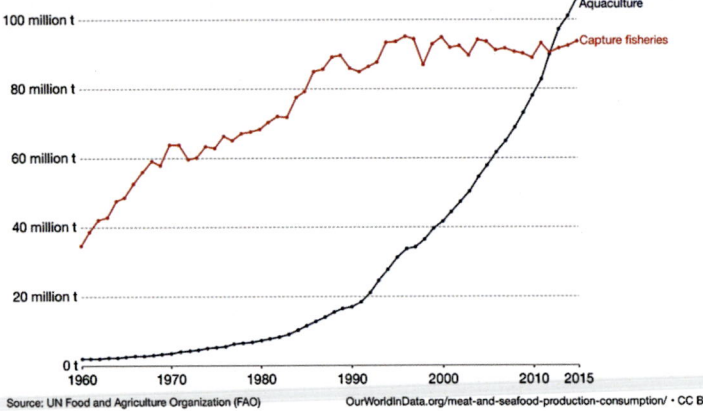

The risks of finfish aquaculture include (i) point-source discharge of untreated fish feces (containing nitrogen and phosphorous—and we know how much damage these elements can do), (ii) release of excess fish food, hormones, antibiotics, chemical food additives, and pharmaceuticals, (iii) further depletion of wild fish stocks due to reliance upon wild-caught anchovies and sardines to make fish food, (iv) spread of disease and sea lice from farmed fish to wild fish stocks, and (v) unforeseen consequences when farmed fish escape and breed or compete with wild fish. Breeding pens can be perfect environments for parasites and diseases that cannot be confined in a fluid environment, such as the infectious hematopoietic necrosis virus and viral hemorrhagic septicemia found in Washington's salmon farms. Farmed fish can escape, allowing them to compete with native wild species for food and territory. In the mid-1990s, more than 613,000 Atlantic salmon escaped from Washington salmon farms and are now deemed in Washington waters to be an invasive species. California bans open-ocean salmon farming, and Alaska bans all finfish farming.

By way of example, let's say there are several thousand finfish of exactly the same species swimming side by side in a circular pen or octagonal pod made of netting that has a 50-foot diameter and is submerged in a bay near you. Now imagine these same fish eating, pooping, swimming, and resting in the same space. With so many fish concentrated in a small area, the disproportionate amount of fish feces, uneaten food, antibiotics, other bodily products, and chemical supplements being released directly into the ocean is not difficult to envision. Don't forget the sea lice and other parasites that can seriously harm the farmed fish and spread to wild fish. When so many animals are confined in unnaturally close quarters, diseases spread easily and quickly. Such operations are essentially industrial finfish feedlots, frighteningly similar to the livestock concentrated animal feeding operations (CAFOs) we discussed in the previous section. I should make a distinction here—the farming of oysters, scallops, clams, mussels, and other bivalves that spend their entire lives filtering organic particulates from the water, does not harm the marine environment. Instead, these shellfish can be used as filtering systems in the restoration of damaged aquatic ecosystems. Consequently, I refer to finfish aquaculture to distinguish it from crustacean aquaculture.

The good news is that aquaculture technology has evolved. The second-generation aquaculture systems feature land-based recirculating tanks, referred to as recirculating aquaculture systems, or RAS for short. RAS are closed-loop tank facilities located on land, which allow for continuous filtering and cleaning of the water. They do not need to be located in or near marine waters, so there is little risk of fish escaping or pollution of marine waters. The second-generation RAS operations eliminate almost all of the risks posed by offshore and nearshore open systems. In these closed systems, biofilters break down fish urine with the help of ammonia-consuming bacteria, and mechanical filters remove fecal material and leftover food. The solid waste gets disposed of elsewhere and can be repurposed as agricultural fertilizer or for biofuel production. A small amount of wastewater exits the system into the environment through tank overflow and filter cleaning—but this water is mostly free of contaminants since it has been treated by multiple processes.

One of the second-generation systems that has been in operation for a long time is that of Blue Ridge Aquaculture, a company raising tilapia in recirculating tanks. It provides a healthy environment for its fish by mechanical filtration, biofiltration, and ultraviolet germicidal light. The system also removes carbon dioxide and ensures that the returning water is adequately oxygenated. Blue Ridge claims to be the world's largest producer of tilapia, producing four million pounds of tilapia per year and shipping between 10,000 and 20,000 pounds of live tilapia every day. According to the company, "These fish are raised without the use of antibiotics or hormones and are free of mercury (undetectable levels from independent studies) and other industrial pollutants"[97] Blue Ridge's technology was featured in an article on aquaculture in *National Geographic* in 2014.

The flow chart below is from Ideal Fish Company, another second, perhaps third generation RAS fish farm. Ideal Fish recovers the fish waste by-products and use it to grow vegetables and other fresh produce in their indoor aquaponics system. The nutrient-rich wastewaters from the fish are perfect for food crops, as the nitrogen and phosphorus from the fish waste are in forms that can be taken up by the plants. Some aquaculture systems are built to generate the power needed to operate the system. Fish farmers are proactively trying to develop and employ sustainable practices, as the promise of farmed fish as an alternative protein source is potentially very lucrative. The push is on to refine these techniques and develop systems that operate at scale.

RECIRCULATING TANK AQUACULTURE

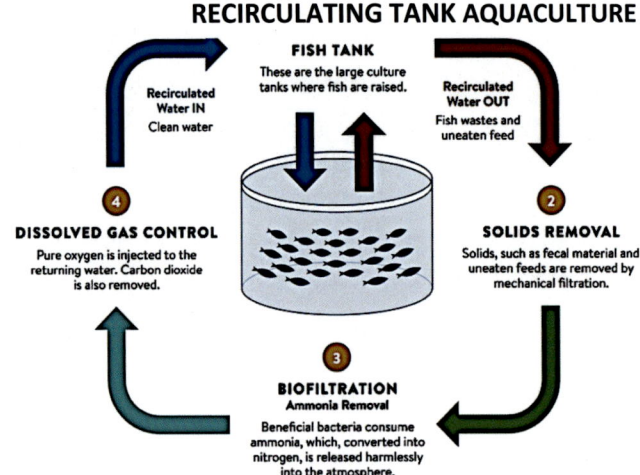

Image from http://idealfish.com/recirculating-aquaculture-ras/

I will end this section with a quote attributed to Jacques Cousteau: "We must plant the sea and herd its animals . . . using the sea as farmers instead of hunters. That is what civilization is all about—farming replacing hunting."

FOOD CROP AGRICULTURE

SUGAR: Lest you conclude that the writer is biased against meat and dairy, let me tell you a story about sugar that is not very sweet. This story demonstrates that non-animal products can also cause abundant harm to our environment. Here is a story about the damage that large-scale food crops can do.

The Everglades is a wetland preserve located at the southern tip of the Florida peninsula, historically thick with birds, mammals, and reptiles. This wetland was created over 6000 years ago after the last ice age, as the receding ocean exposed a bare limestone plain covering much of South Florida. The eastern and western edges of this broad plain can be seen on satellite imagery as Florida's continental shelf. As rain fell in Central Florida, the Kissimmee River Basin was formed, which terminated at Lake Okeechobee, a sizeable lake in the southern third of the state and the situs of Florida's current sugar farming industry. Historically, during the summer wet season, Lake Okeechobee would overflow its southern shore, with its waters flowing slowly south to Florida Bay at the tip of the peninsula. The fresh water traveled slowly through thru sawgrass, past islands of mahogany trees, and past wild orchids as it made its way to Florida Bay at the tip of the peninsula. This beautiful, slow-moving "river of grass" once covered most of the southern third of the state. As the water made its way south to Florida Bay, the underlying aquifers were replenished and recharged. A wide variety of animal and bird species thrived for thousands of years in this exquisite and self-sufficient ecosystem. Then the Spanish, British, and Americans showed up, giving birth to a saga of greed, power, and disregard for this unique ecosystem, the only one of its kind on the planet.

At the turn of the twentieth century, the common belief was that the natural environment was man's to tame, to make it work for humankind (or at least for a handful of powerful men). Political power at the federal and state levels created the agencies that would have the authority to reshape the Everglades into an accommodating place for agriculture and housing developments. The process of remaking the Everglades involved (i) eliminating the fertile Kissimmee River Basin by turning the upper river into an arrow-straight canal, (ii) dredging and filling large portions of the wetlands of the Everglades, (iii) expanding the Caloosahatchee River on the west and the St. Lucie River on the east of Lake Okeechobee, so that there was a continuous trench from the Atlantic Ocean to the Gulf of Mexico, and (iv) constructing a complicated plumbing system of pumps, dikes, valves, levees and locks, designed to control water flow over most of the state of Florida. But who had possession of the key to the plumbing fixtures? You already know the answer: politicians with close relationships with banking and real estate development magnates.

After the state of Florida was replumbed, the excess water flowing into Lake Okeechobee from Central Florida was mechanically pumped and directed east to the Atlantic Ocean and west to the Gulf of Mexico by way of the Caloosahatchee and St. Lucie Rivers. Water no longer overflowed the banks of Lake Okeechobee to flow slowly south to Florida Bay and recharge aquifers along the way. Soon the wetlands dried out. After the "draining of the swamp," as this project came to be known, came real estate development and a massive population explosion in South Florida. At the same time, thousands of drained acres of former Everglades land were planted with sugar cane. The Sugar Act of 1934 provided millions of dollars of subsidies to sugar producers. US Sugar was founded in 1931, and in 1954, Florida Crystals was formed by a Cuban sugar baron fleeing his country after Fidel Castro came to power. Along with the sugar cane plantations, came fertilizer use, soil erosion, and fresh water consumption.

You can easily imagine the impact the draining, farming, fertilizing, dredging, filling, construction, and plumbing were having on Lake Okeechobee and the Everglades. Unfortunately, Mother Nature did not get the memo about the repurposing of the Everglades. Today, Florida Bay is undergoing ecological collapse, as the lack of fresh water from Lake Okeechobee has increased the salinity of Florida Bay, killing its seagrass, and causing the entire trophic food web to suffer. Today, the water flowing east to the Atlantic and west to the Gulf is loaded with fertilizer products and other chemicals from farming. When the nutrient-rich water reaches the Atlantic and the

Gulf, the naturally occurring algae enjoy a feeding frenzy as they consume the fertilizer nutrients in the water. The feeding frenzy creates gigantic algae blooms that steal all the oxygen from the near Gulf and Atlantic waters, kill seagrass, fish, and shellfish, and cause breathing distress to humans. Tourist revenues in South Florida were down by as much as 50% after the 2018 red-tide algae blooms, resulting in losses in the hundreds of millions of dollars.

The Everglades Forever Act (EFA) was passed in 1994; it was a well-intentioned cleanup bill, with minimal associated funding. Big Sugar fought it tooth and nail and has continued fighting funding efforts and water-control efforts. With that said, the EFA has produced some improvements. Phosphorus levels are now down, and for the first time, during the summer of 2019, no fertilizer-laden waters were released into the Caloosahatchee and St. Lucie Rivers—the algae blooms that summer were practically nonexistent. South Florida Water Management District, the state organization that runs this show, did not bow to Big Sugar's demands for water drainage; consequently, the residents, marine life, plant and animal life, and tourist industry of Florida benefitted. The federal government has joined the effort to repair this broken ecosystem, creating the Comprehensive Everglades Restoration Plan (CERP), a complex and expensive environmental restoration program. Most scientists agree that the only way to restore this ecosystem is to reestablish the historic southerly flow of fresh water from the Kissimmee River Basin to Lake Okeechobee and south to Florida Bay. As this book goes to print, the reservoir to be constructed directly south of Lake Okeechobee has been approved by the State of Florida. This reservoir is the most significant step in the plans to restore the historic flow of water. I understand that dirt is already moving!

Florida is the number one sugar producer in the United States, accounting for half of all sugarcane acreage and generating between $1.3 and $1.6 billion in total income and over 18,000 full-time jobs. The demand for refined sugar only increases as our population grows. Think about this story the next time you reach for a sugary treat.

OTHER PLANT FOOD CROPS: As we approach a world population of 9 billion by 2050 on a planet with diminishing land, degraded and limited water resources, disrupted biogeochemical cycles, climate extremes, worn-out soils, and industrial monoculture farms, how do we ensure a safe, affordable, nutritious food supply for everyone? "If these [sic: planetary] boundaries were strictly respected, the present food system could provide a balanced diet (2,355 kcal per capita per day) for 3.4 billion people only."[98] Not only is the current agriculture footprint too large for sustainability, but it also lacks the headroom needed to expand to meet growing demand. Please note that the term "agriculture" technically refers to both the growing of plants and the breeding and raising of livestock, even though common parlance sometimes limits the term "agriculture" to the growing of crops and other plants. So, when you hear statistics quoted about agriculture, be sure you understand what is intended to be included.

After World War II ended and soldiers came home from the Pacific and European fronts, there was a boom of babies born over the next 18 plus years. At the same time, there was a green revolution as the country attempted to meet the demands of this swelling population. To meet this increasing demand, (i) the size of individual farms increased, (ii) new technologies and equipment were adopted, (iii) chemical fertilizers, herbicides, and pesticides were improved, and (iv) new varieties of rice, corn, and wheat were developed. These advances spread around the world in the 1950s and 1960s, and farmers saw their yields increasing. Since the green revolution in farming, we have shifted to a system that produces most of our food through industrial agriculture, composed of large corporate-owned farms growing the same crops each year with large amounts of fertilizer, pesticides, and herbicides. As a result, our topsoil is depleted of nutrients, surface waters and groundwaters are contaminated, we are emitting considerable amounts of greenhouse gases, family farms are disappearing, and the economies of our rural communities are suffering. This system is not sustainable, and we must learn to produce our food without harming the natural environment upon which the growth of that food depends.

We explored in Chapters 2 and 3 the workings of our historically balanced, renewable biogeochemical cycles. Applying that knowledge, you can see how industrialized food crop agriculture has become disconnected from our ecological cycles in the minds of consumers. Historically, food crop agriculture was a balanced ecosystem, with family farms owning a few animals which they grazed and fed with a portion of the food crops they grew, while using the waste from the animals to fertilize the crops. The health and fertility of the soil were critical to the survival of the farm and its occupants. Unless farmers made sure that their agricultural practices nourished the soil, their families would not eat. Modern agriculture techniques developed over the past 75 years have prevailed over the sustainable methods of the past. Bringing back sustainable crop production methods, improved by today's technology, can restore this ecosystem.

As discussed earlier, a cup of fertile soil should contain thousands of species of bacteria and fungi, along with hundreds of species of nematodes and protozoa. Remember atmospheric nitrogen and reactive nitrogen? During the slow-growing season, farmers frequently plant cover crops like oats, rye, or clover, which, at maturity, are plowed under to provide organic matter, improve soil structure, and reduce erosion. Other sustainable practices include no-till or low-till practices. We learned that soil is one of our carbon sinks, so you can imagine the carbon released into the atmosphere when a large field is mechanically tilled. Instead, if the plant waste remaining after harvest remains on the field, it can provide nutrients for the soil. Another element of sustainable crop farming involves plant diversity. Monoculture farming increases the need for fertilizer, pesticides, and herbicides and exhausts the soil's nutrients more quickly. We mentioned earlier that we are losing millions of acres of cultivated farmland per year to soil erosion. Similarly, crop rotation (moving a given food crop from one field to another) helps reduce the need for fertilizer, because various crops leave behind distinct nutrients and create different soil biomes.

In 2018, representatives of more than 70 countries gathered in Rome to discuss, once again, how to achieve a more sustainable agriculture system. José Graziano da Silva, the director-general of the United Nations FAO, called for "transformative change toward sustainable agriculture and food systems based on agroecology." In addition, the economics of food production are starting to change, incorporating what is being referred to as "true-cost accounting" to assign values to industrial agriculture's contribution to soil erosion, climate change, and public health. We will discuss this in more detail in the final chapter.

BOTTOM LINE: FOOD. Who knew that what we chose to eat could have such a powerful impact on the health of our planet? The meat and dairy industries win a triple crown for achieving all of the 3 Ds of harm to our planet: depletion of nonrenewable resources, degradation of renewable resources, and disruption of natural biogeochemical cycles. After the cost of the harm from the 3 Ds, we have to look at the cost of the damage that our consumption of animal products can cause to our bodies. A 2018 report from the Milken Institute estimated US health care costs for chronic lifestyle diseases (e.g., heart disease, obesity, diabetes, and cancer) totaled $1.1 *trillion* in 2016, and when lost economic productivity was also considered, the total economic impact was more like $3.7 trillion.

And to add insult to injury, you and I are paying for the meat and dairy subsidies that help make this all possible. Finally, consider this: humans are the only animal species on the planet who drink the milk from the mammary glands of a different animal species, a definite yuck factor. If you read *The China Study* by T. Collin Campbell, you can learn about his well-documented research covering the cancer-producing characteristics of cow's milk products when consumed by humans.[99] The almond, soy, and other milk substitutes are delicious, affordable, and enticing a significant share of the US milk market away from traditional dairies. In November 2019, a large milk producer in the United States announced it was filing for bankruptcy, citing reduced consumer demand for dairy products as one of the causes. Will this happen with meat? Will we find a sustainable way to farm fish? Germany is experiencing a strong vegan/vegetarian movement as significant portions of its population give up meat and dairy. In response, German meat producers are starting to produce and sell vegetarian and vegan alternative-meat products. Also responding to increased demand, US grocery shelves and restaurants are becoming stocked with an increasing variety of non-meat choices.

I will close this chapter with a restated comment from the FAO, which provides sobering food for thought and should prompt us to think about population growth momentum and overpopulation.

World average per capita availability of food for direct human consumption, after allowing for waste, animal feed and non-food uses, reached 2,770 kcal/person/day in 2005 to 2007. Thus, in principle, there should be sufficient aggregate global food for nearly everyone to be well-fed. Yet this has not happened: some 2.3 billion people live in areas that can barely provide 2,500 kcal[person/day], and some 0.5 billion people live in areas struggling to provide 2,000 kcal[person/day], while at the other extreme some 1.9 billion people live in countries that are able to provide *more* than *3,000* kcal[person/day].[100]

Chapter 6

TOWARD SUSTAINABILITY

Some of the green technology solutions that can help bring this planet back into balance already exist. Much of this technology has not been scaled, and much of it is not ready for prime time. The taxpayers of the United States dish out $38 billion per year for livestock and dairy subsidies, $649 billion per year for oil subsidies, and $4 billion per year for sugar subsidies. These amounts have to be considered approximations, as they are difficult to track and quantify. They come in many forms other than direct cash payments, such as loan guarantees, insurance, marketing loans, price supports (guaranteed prices), import quotas, and tax concessions, all of which reduce the cost of doing business. Just imagine what these same dollars could do for research, development, and funding for start-up companies.

Many of these start-up companies are playing beat-the-clock to refine and scale their technologies before we exceed our biocapacity and find ourselves with no return ramp. Venture capital opportunities abound, as our best and brightest struggle to develop workable solutions to the many problems outlined in the previous chapters. I am going to highlight a few of the technological solutions I have found particularly interesting. I almost did not include this chapter, as I do not want you to infer that we can rely on technology to bail us out of this predicament. In the final analysis, technology cannot transform finite resources into infinite resources. On the other hand, certain technologies might be able to stretch our resources sufficiently to buy us time to bring our consumption patterns in line with the limitations of our planet. Still, we must not forget that the long-term solution lies with us—with our food choices, family planning decisions, and consumption choices. It also lies with our strident, collective voices, which have the power to bring essential change at the policy level.

What are green technology solutions? I watched a TED talk by Dr. Bilal Bomai of NASA's Green Lab, and his definition stuck with me. This definition also appeared in a 2011 article by Dr. Bomai and read as follows:

> To be truly labeled a green solution, three conditions must be met. The solution must be alternative, renewable, and sustainable. (i) An alternative solution is one that is not in mainstream use today and has no undesired consequences when compared to conventional solutions, (ii) A renewable solution is derived from sources that are naturally replenished, and (iii) A sustainable solution refers to the ability to maintain ecological processes, functions, biodiversity, and productivity into the future. For any technology, technique, or method to be considered green, all three conditions (alternative, renewable, and sustainable) must be met inclusively.[101]

AVIATION BIOFUEL: The global aviation sector uses more than 80 *billion* gallons of fossil-based fuel per year.[102] I could not resist the temptation to expand the math—if aviation uses 80 billion gallons of fuel per year, that amounts to 152,000 gallons per minute, or 2,500 gallons per second. Shifting to renewable aviation fuel sources is a massive challenge if existing aviation infrastructure is to be used. As Tony Radich states in his paper presented at the 2013 biofuels workshop held by the U.S. Energy Information Administration's Office of Petroleum, Natural Gas, and Biofuels Analysis.

> Aircraft and airport fuel storage and delivery systems are designed to last for decades; new fuels must be compatible with existing systems. Non-petroleum jet fuels such as biojet

must consist entirely of hydrocarbon compounds that are already found in petroleum jet fuel. In other words, biojet must be a drop-in biofuel. The approval process for new formulations of jet fuel is very involved, due to the range of conditions under which jet fuel must perform. A plane may take off from a scorching Arizona desert, climb to a freezing 30,000 feet, and land in a humid Louisiana swamp. Under all these varied conditions, the fuel can't freeze, boil, or absorb water.[103]

Researchers at the US Department of Energy's Lawrence Berkeley National Laboratory are using carbohydrate-rich plant material, along with genetically modified bacteria, to digest the plant sugars into energy-dense molecules that can be chemically converted into a fuel product. NASA's Green Lab is working with salt-tolerant plants called halophytes to make jet biofuel in closed bioreactors. In 2010, a test flight of the US Navy's F/A-18 "Green Hornet" aircraft was powered by camelina plant-based biojet fuel at the Naval Air Warfare Center Aircraft Division in Maryland. The flight was powered by a 50-50 blend of camelina-derived biojet fuel and traditional petroleum-based jet fuel supplied by Honeywell UOP. In light of the prodigious fuel consumption by aviation, this potential 50% reduction in petroleum-based consumption could be a significant interim step.

OTHER BIOFUELS: Two biofuels currently in use on our highways are corn-based ethanol and biodiesel. In the United States, ethanol production from corn crops began in the early 1970s after an OPEC oil embargo left many countries high and dry. The complaints about ethanol from corn crops are that corn ethanol competes with food crops, corn is not an efficient fuel source, and that high production costs render corn ethanol an impractical substitute. As the technology has improved, biofuel production has moved to inedible cellulose material (e.g., agricultural waste) rather than food crops. Second and third-generation biofuels are currently a collection of technologies at various stages of commercial development. Biofuels from soy, jatropha plants, sugarcane, sorghum, and algae are being developed. ExxonMobil is working on an algae-based biofuel. The company has partnered with Synthetic Genomics to double the fat content of algae by genetic modification. More fat means that biofuel can be produced more efficiently.

As improved second and third-generation biofuels are being developed, there have been stumbles from technical failures, cost overruns, and investor pressures, as companies have attempted to scale up production. The 2013 grand opening of the third commercial-scale cellulose-based ethanol biorefinery in the United States was notable on a couple of levels. First, the raw material is corn crop debris harvested from local fields, so there is no competition with food crops and minimal transport costs. Secondly, this refinery was the result of a public-private partnership with the Department of Energy's Office of Energy Efficiency and Renewable Energy (EERE). This facility, Abengoa Bioenergy Biomass of Kansas, uses a proprietary process to turn nonedible corn stalks, stems, and leaves, which are locally harvested, into fermentable sugars that are then converted into transportation fuels. The plant is contemplating adding an electricity cogeneration component that will generate up to 21 megawatts of electricity, which should provide enough alternative energy to power the plant and the local community.

Government involvement appears to be critical for the development of advanced biofuels. Industry insiders believe that one of the reasons for the sluggish progress in bringing advanced biofuels to market is uncertainty and difficulties with investor funding. Redirected government subsidies could be part of the solution. A 2019 article from the International Monetary Fund reported that the global community spent $5.2 *trillion* on fossil fuel subsidies in 2017, 40% of which was directed to the *coal* industry.[104] Really, we are subsidizing *coal*? Taxpayers from all over the world actually spent $2 *trillion* to promote a fuel that emits sulfur, mercury, lead, and arsenic, along with copious amounts of carbon. I thought subsidies were intended to encourage the subsidized activity.

MEAT SUBSTITUTES AND CLEAN MEAT: As our global population approaches Earth's carrying capacity, food and water insecurity will continue to increase. According to the United Nations FAO, global demand for meat is expected to increase by 70% by 2050. However, we know from Chapter 5 that current meat production methods are not environmentally sustainable, and there is not enough arable land to grow food for an expanded livestock population. After reading about the damage that the livestock industry inflicts upon our planet, it is easy to declare that everyone needs to stop eating meat. That notion is unrealistic on many levels.

We currently have plant-based alternatives to meat products. These products emulate meat, and while they may be delicious, they do not come close to duplicating the taste, mouthfeel, texture, and cooking behavior of meat. Some of these products are starting to get closer to mimicking animal products, like JUST Egg, Beyond Meat, and almond milk. The veggie burger produced by Impossible Foods contains a soy product known as leghemoglobin, which, when cooked, produces an iron-rich compound called heme. Heme lends a meat-like taste and appearance to the plant-based burgers. Any plant-based alternative product will need to be nutritionally dense in its own right and free of unhealthy ingredients, such as excess sodium. Simply mimicking animal products will not, in the long run, inspire a meat-eating population to switch to plant-based meat products.

On the other hand, on the near horizon is clean meat, also called cultured meat. Cultured meat is actual animal tissue grown from stem cells from a donor animal. What is a chicken breast other than a group of chicken muscle and fat cells held together by the chicken's connective tissue? Why do animal cells have to grow in a living animal that will have to be killed? In making cultured meat, a small number of muscle stem cells are taken from the donor animal while under anesthesia. These are muscle stem cells that typically lie dormant in animals (including you and me) until a severe muscle injury occurs, at which time these cells are called upon to repair and build new muscle tissue in the living animal. These muscle stem cells are transferred to a bath of growth factors and nutrients (sugars, amino acids, fats) similar to the nutrients which feed the cells inside a growing animal. In this bath, the stem cells are allowed to reproduce and grow until there are trillions of cells. At some point, the cells begin to differentiate into muscle cells where they bulk up and start to resemble the strands and texture that are present in meat. The growth of cells in the nutrient bath is essentially the same process that would occur inside the live animal. One biopsy sample can make tens of thousands of hamburgers or boneless chicken breasts.

The first cultured hamburger was produced by MosaMeats, cost $300,000, and was presented at a press conference in 2013. Sergey Brin, a cofounder of Google, provided a significant portion of the funding for the research and development that produced this hamburger. Numerous other companies are working to develop similar products. The pricing has come down, and the focus has been on flavor, texture, and scaling for affordable mass production. Companies like Tyson Foods and NGOs like the Gates Foundation have invested in this technology.

This technology is full of advantages: It is real meat, produced in a sterile environment (think back to manure management issues in Chapter 5), no animals are killed, no cruel conditions are involved, no antibiotics are needed, no growth hormones are used, and no genes are modified. The best part of cultured meat is that some of the land that is being used to grow food for animals could be converted to grow plants for humans. With more arable land available, farmers can allow some land to lie fallow, can rotate crops, can get away from monocrop agriculture, and can better manage the fertility of the soil. The very same companies that are invested in the livestock industry infrastructure can be invested in the production of cultured meat, as demonstrated by the example of Tyson Food's 2018 investment in Israel's Future Meat Technologies.

Like the advanced biofuel start-ups, these young cultured-meat companies are struggling to secure investors and face the same technological challenges of scaling for market that the biofuel companies face. Another challenge to funding comes in the form of public awareness of the need for meat alternatives. The public is much more aware of the need for renewable energy sources than they are aware of the damage done to our environment by the livestock industry, so speak up when the opportunity arises.

THORIUM NUCLEAR ENERGY: Now, here is a development that is very exciting and is actually not new at all. Thorium is a radioactive element fully capable of fueling nuclear reactors, in the same way that uranium does, but it became the underdog fuel during World War II. During the war, the United States was exploring the development of an atomic bomb through the Manhattan Project. The Manhattan Project nuclear research program was laser-focused on developing a bomb, not alternative energy sources. Uranium was considered the choice element because it is capable of producing plutonium, critical for making bombs. After the war, the United States continued its fixation with nuclear weapons and moved forward with uranium as the energy fuel of choice. As a result, all of our power plant infrastructure was designed around our knowledge of the chemical properties and behavior of uranium.

Have you wondered how conventional uranium nuclear reactors produce energy? We will review the operation of conventional reactors so that we can distinguish thorium reactors. Radioactive elements such as uranium and thorium are heavy, with a large number of neutrons and protons in their nuclei. There are three naturally occurring forms of uranium: U_{235}, U_{234}, and U_{238}, and they differ as to the number of protons in their respective nuclei. The extremely slow decay of unstable uranium atoms is the mechanism that heats the inner and outer cores of our planet, drives the convection in Earth's outer core and mantle, and ultimately moves our landmasses. The strong nuclear force holds together the protons and neutrons in the nucleus of each uranium atom. This force is much stronger than the chemical bonds in any of our fossil fuels, and therefore has much more potential energy to give up.

In most conventional reactors, uranium$_{238}$ is used, along with a small amount of U_{235}. When a uranium atom is hit by a single neutron, the nucleus of the atom splits, produces heat energy, and throws off some of its neutrons. Those neutrons thrown off then bombard other uranium atoms, causing those nuclei to split and to throw off some of their neutrons, producing a chain reaction. The breaking apart of the nuclei releases the potential energy of the strong nuclear force. The energy creates enough heat to boil water that produces steam that moves a turbine that generates electricity. The entire nuclear reactor is bathed in pressurized water to moderate the speed of the chain reactions. Because of the pressurization, the entire system must be housed in a large containment structure. There are also control rods filled with materials (e.g., boron, cadmium) that can slow down the chain reaction when necessary by reducing the number of excess neutrons available. The waste products, once the uranium is exhausted, are U_{238}, U_{235}, and plutonium, which take thousands of years to break down, and therefore, must be safely stored. As the world has witnessed too often, there can be safety issues with uranium-powered nuclear reactors. That was an oversimplified explanation, but it will be useful in examining how thorium differs.

By comparison, thorium is a heavy metal like uranium and comes from the same neighborhood on the periodic chart as uranium, but has some unique and desirable characteristics. Thorium is far more abundant in nature than uranium, and it naturally occurs in higher concentrations than uranium. As a result of these characteristics, thorium is easier and less expensive to extract. In addition, thorium is not very reactive on its own (not very fissile), does not have to be enriched like uranium, produces less waste than uranium, is not very useful for weapons use, and generates more energy per ton. Finally, because of the technology employed, thorium reactors are less dangerous to operate.

There are several thorium reactor technologies being studied, and the following is a description of a molten salt reactor. This type of thorium reactor starts with a form of thorium designated as Th_{232} and a bit of uranium to initiate the reaction, with the thorium nuclei splitting, throwing off neutrons, and producing energy in a chain reaction, as with a uranium reactor. The Th_{232} becomes Th_{233} when it is struck by and absorbs a stray neutron, and because Th_{233} is highly unstable, the chain reaction continues on its own. The heat from this reaction is transferred to a loop of molten salts that do not contain any fissile material. Heat energy is then transferred to helium gas, which turns a turbine, which generates electricity. The reactor is not pressurized and is self-regulating because the thorium will simply become solid again if power is lost. In the event of some other emergency, there is a plug that melts, allowing the salts to drain into a pan. There is no leftover thorium, just a small amount of U_{235}, which can be used over and over, along with all of the U_{235} currently in storage. Since thorium is not useful as a weapon, there is virtually no risk if every country on the globe were to adopt thorium as a fuel for nuclear reactors. While expensive to construct, thorium reactors are less costly to operate than conventional uranium reactors, can be produced as modular units, and can be scaled up or down for a specific application. In addition, the small amount of waste produced by thorium reactors breaks down in 300 to 500 years. Finally, thorium reactors should prove to be a much, much safer nuclear energy alternative.

Both China and India are working on the development of thorium reactors, with several different kinds of technologies being considered. China was working with the Bill and Melinda Gates Foundation to build a prototype molten salt thorium reactor, but the project was put on hold when trade negotiations with the United States broke down in 2019. Thorium reactors involve technology that is more likely to be brought to market by government or public-private partnerships, as it is probably too large a project to be managed by private investment alone. Because of the vested interests of US power companies and the US Atomic Commission in the older uranium power technology, bringing thorium to the head of the class is going to be a challenge in the United States. Just stop for a moment to consider the paltry amount of energy we derive when we break the hydrogen-carbon bonds of fossil fuels by combustion, compared with the significant amount of energy derived from splitting the nucleus of a heavy element like thorium. Add to the mix, the scalability and increased safety of thorium reactors, and thorium reactors begin to look like a very attractive alternative energy source. We need to speak up to stop the negative collective Pavlovian response to nuclear energy.

FERTILIZER: You may be surprised that among these various high tech innovations described in this chapter, I would mention fertilizer. Technology has fostered rapid growth in crop productivity, particularly during the second half of the twentieth century. None of the new technology has had a more dramatic impact on crop yields than synthetic nitrogen fertilizer. In addition to water and sunlight, crops need three key nutrients to grow: nitrogen, phosphorous, and potassium. Nitrogen is often the nutrient that is most limiting to increasing crop yields, despite more than 78% of Earth's atmosphere consisting of it. In the atmosphere, nitrogen exists in the form of N_2, rather than as nitrates, which is a form of nitrogen that plants can absorb (take a look back at the nitrogen cycle in Chapter 2).

For thousands of years, populations had to rely on recycling the limited quantity of usable nitrates naturally occurring in soils, along with that which could be found in animal waste and other biomass. Using these naturally occurring sources of nitrates meant that existing crops could only support a limited population. In 1881, Chile invaded Peru, seeking to capitalize on Peru's extensive guano accumulated on several of the barrier islands along Peru's coast. With quickly growing populations, the bird guano soon was depleted, meaning that the population was again facing a global carrying capacity crisis. This is classic Thomas Malthus once again.

I mentioned in Chapter 2 that in the early 1900s, a German chemist named Fritz Haber developed a process to convert atmospheric nitrogen into ammonia. Another German chemist, Carl Bosch, was able to scale Haber's process to an industrial level. This was a spectacular development, creating nitrogen fertilizer out of thin air! Some commentators consider the Haber-Bosch process as one of the most significant technological advancements of the twentieth century. The process uses extremely high heat and pressure to force a chemical reaction between atmospheric nitrogen and hydrogen from natural gas. The process must use these extreme conditions because nitrogen gas molecules are held together with strong triple bonds. Is it possible that the Haber-Bosch process prevented our population from reaching a Malthusian limit during the postwar baby boom, and most of us just unwittingly skipped by it?

Pivot Bio, with financial support of $70 million from Bill Gates, Jeff Bezos, Michael Bloomberg, and Richard Branson, is developing nitrogen-fixing bacteria to enrich our tired, overused soils. You may recall that specific bacteria in the soil convert atmospheric nitrogen to a form of nitrogen that can be used by plants. The Pivot Bio concept is to provide farmers with bacteria "seeds," which farmers can deposit in the soil, along with a probiotic to activate the bacteria. The activated bacteria seeds soon start producing nitrogen in a form that is usable by plants. Two other start-ups, Azotic Technologies based in England, and Intrinsyx Bio in Silicon Valley, are working on similar technology with another type of bacteria. These innovations could be the most significant environmental breakthrough in agriculture since Haber-Bosch.

Shifting to phosphorus, I mentioned earlier the discovery of phosphorus by Hennig Brand by evaporation and condensation of buckets of human urine. Phosphorus is another limiting nutrient, like nitrogen. I described how we extract phosphorus from the earth by strip-mining phosphates from rock. Globally, we have about 70 to 100 years of economically accessible phosphorus reserves remaining (USGS estimates are closer to 260 years). In either case, it is a critical element for all life and for ensuring that we can feed our increasing population. Consequently, current technology has us back in the toilet, searching for solutions.

In 2011, Aarhus Water, a municipal water treatment company in Aarhus, Denmark, began recovering phosphorus from wastewater in partnership with pump manufacturer Grundfos. The process involves running a stream of wastewater through a bioreactor, adding magnesium salt, then allowing the sludge to precipitate out. The process refines the phosphorus in the wastewater and discards heavy metals and other environmentally unfriendly substances. The end product is a granulate that contains phosphorus in a form perfect for use as fertilizer. The system is not scaled to meet the demand of Denmark's agricultural sector, which annually requires 11,000 tons of phosphorus. The company's spokesperson indicated that if similar recovery plants were built at the 50 largest water treatment plants in Denmark, Aarhus Water could produce 3,000 tons of pure phosphorus fertilizer per year, representing more than 25% of the country's needs. In researching this subject, I learned that there are multiple technologies being developed to recover and recycle phosphorus from wastewater. A 2009 article published by Water Science and Technology compares several very different technologies for extracting phosphorus. [105] Phosphorus can be recovered from wastewater, sewage sludge, as well as from the ash of incinerated sewage sludge. The phosphorus recovery rate from the liquid phase can reach 40% to 50% at the most, while recovery rates from sewage sludge and sewage sludge ash can reach up to 90%. While there are various methods that can be applied for phosphorus recovery, until now, there has been limited industrial-scale implementation.

Food demand is on the rise globally. There is an increasing demand for meat and dairy, especially in rapidly growing economies. This demand means there is and will continue to be an increasing need for nitrates and phosphates to use as fertilizer. Most of the phosphates and nitrates used in farming today are lost during the process of fertilizer production, excessive application on fields, uptake by crops, post-harvest crop waste, and food waste. A July 2018 article by Dr. Hisao Ohtake, professor Emeritus at Osaka University, published in the International Water Association Journal, reported that there are more than 70 full-scale phosphorus recovery plants operating in Europe, North America, and East Asia. In the same article, Dr. Ohtake reported that phosphorus recovery has recently become an issue in the world of high tech manufacturing. High-purity phosphorus compounds are required for the manufacture of certain technical products, including semiconductors, lithium batteries, liquid crystal panels, and fire-retardant plastics. Apparently, pure elemental phosphorus is becoming increasingly scarce, produced by only a few countries by a labor- and energy-intensive process. I hope that this demand from the high tech sector will bleed into the agriculture sector, infusing more funds and interest in producing scalable, cost-effective phosphorus recovery methods.[106]

VERTICAL FARMS: With an estimated population of 9.7 billion people by 2050, approximately 70% of which will be living in urban settings, food security is a looming issue. Our planet has, at best, a fixed amount of arable land that can be employed to produce food for our growing population. Compounding food security concerns are the environmental harms from livestock agriculture and food crop agriculture, which we have discussed in detail. One very interesting solution to these challenges is the vertical farm. Vertical farms grow plants in vertically stacked layers integrated into multistory buildings under artificial light. Crops stacked on top of each other can be cultivated to maturity more quickly than with traditional farming methods. Think of a vertical farm as a high-rise greenhouse on steroids. Vertical farming typically operates in a controlled environment in which temperature, light, and humidity are, for the most part, artificially controlled. Instead of soil, either water or other organic materials like coconut husks are used as a growing medium.

The advantages of vertical farming are numerous: (i) vertical farms can be up to 100 times more productive than fields, (ii) the plants can be grown without pesticides or herbicides because of the contained environment, (iii) the plants use 75% to 95% less water, (iv) the growing season is 12 months a year, and (v) the farms can provide fresh produce in close proximity to growing urban populations. The primary drawbacks are (i) high upfront and operating costs, (ii) dependency upon technology, (iii) operating energy requirements and costs, (iv) the limited kinds of food crops that can be grown in this manner, and (v) pollination challenges. Even so, I find this technology very appealing, and apparently others have as well. According to an April 2019 article in Forbes magazine, this technology has attracted venture capital from numerous sources, including Softbank Group ($200 million), the Sheik of Dubai with IKEA ($115 million), and Google Ventures ($90 million). I am not alone in my interest in this new industry. The author of this Forbes article, Erik Kobayashi-Solomon, probed more deeply to unearth the good, bad, and ugly of vertical farming. His conclusions were as follows:

- Microgreens, herbs, and lettuces grow well in these environments, but other crops may not be as compatible with the vertical farm environment. Market demand will not be satisfied with just leafy greens for long.

- We learned in Chapter 4 that *Homo sapiens* has been farming for over 11,000 years, but the technology and know-how accumulated over those years are not necessarily transferrable to vertical farming, an unnatural habitat for growing food. This is going to take research and time.

- The largest operating expense of vertical farming is the lighting, and even with the use of LED lights, this factor must be overcome for vertical farms to operate profitably.

- So long as the costs associated with the environmental damage caused by our current agricultural system are not factored into the cost of the final products we purchase, produce from vertical farms will never be competitive with field-grown produce.

- Vertical farms, despite their high capital and operating costs, may be a sound solution in sunny, water-scarce areas such as the Middle East, where most produce must be imported.[107]

The 2018 US Farm Bill established the USDA Office of Urban Agriculture and Innovative Production to encourage emerging agricultural production practices, including the authority to provide grants to support research of alternative food technologies. In April 2020, the USDA announced the availability of $3 million for grants. While the 2018 farm bill did authorize funds to support alternative food technologies, the USDA has continued subsidies to companies using the industrial farming methods we previously discussed. These continued subsidies will make it impossible for vertical farms and other alternative food technologies to compete with industrially grown crops. If you are interested in learning more about this technology, go to http://www.80acresfarms.com to learn about their operation and products. The image below is from the USDA website and depicts what a vertical farm might look like.

credit: Oasis Biotech, from USDA website, according to terms of use

CARBON DIOXIDE CAPTURE: Technology has been developed to remove excess carbon dioxide from the atmosphere and store the excess carbon dioxide underground. There are two groups of carbon dioxide capture technologies: (i) the type that captures carbon dioxide emissions on-site as part of a fossil fuel combustion facility and then stores it underground (CCS) and (ii) the type that captures carbon dioxide emissions from ambient air and then stores it or uses it to make fuel (DAC). The former is already in practice, but the latter is in the early stages of development.

Carbon dioxide capture and sequestration (CCS) could play a significant role in reducing greenhouse gas emissions from one of the largest emissions sources. According to the United States EPA, more than 40% of carbon dioxide emissions in the United States are from electric power generation. The EPA has estimated that CCS can reduce carbon dioxide emissions from fossil-fuel-operated power plants, by 80-90%. The EPA compares that reduction in carbon dioxide to planting more than 62 *million* trees, and then waiting at least ten years for them to grow. The EPA suggests that CCS could also be used to reduce emissions from industrial processes such as cement production and natural gas processing facilities. The process involves three basic steps: (i) capture of carbon dioxide emitted from power plants or industrial processes, (ii) transport of the captured and compressed carbon dioxide by pipeline, and finally, (iii) underground injection and geologic sequestration of the carbon dioxide in deep underground rock formations, such as depleted gas reservoirs, un-mineable coal seams, shale rock, and basalt formations. Overlying these formations are impermeable, non-porous layers of rock that trap the carbon dioxide and prevent it from migrating upward. On the following page is a graphic from the EPA depicting the CCS process.

In a recent article by Anuradha Varanasi of Columbia University Earth Institute, she stated that there are 43 commercial CCS facilities all over the world, some in operation and others in varying stages of development. Ms. Varanasi cites the International Energy Agency, which reported that globally more than 30 million tons of carbon dioxide are being captured from large-scale CCS facilities every year. Over 70% of this is happening in North America. Ms. Varanasi also cites Julio Friedmann, a senior research scholar at the Center on Global Energy Policy at Columbia University, who asserts that the technology is reducing carbon dioxide emissions by 55% to 90%, is removing other pollutants like nitrous oxide, and is now costing $40 per ton of carbon dioxide. The article states that there is widespread consensus among scientists and geologists that the storage step of this process is safe. In some of the facilities, the captured carbon dioxide is injected directly into oil reserves to make it easier to extract the oil. The carbon dioxide is then retained in the layers of rock that previously held the oil.[108]

CARBON CAPTURE AND SEQUESTRATION

From the EPA according to terms of use

Turning to the capture of carbon dioxide directly from ambient air, we find that there is a fundamental difference between direct air capture (DAC) and conventional CCS described above. DAC can remove excess carbon dioxide from the atmosphere independent of its source, while conventional CCS requires a concentrated source (i.e., flue gas from a coal-fired power plant). While both processes are important technologies to reduce atmospheric carbon dioxide, the primary advantage of DAC is that it is not dependent upon fossil fuel energy production, and can continue capturing atmospheric carbon dioxide long after we have moved away from fossil fuel energy sources. The most worrying challenge to DAC is the energy cost entailed in operating the units.

There are two kinds of DAC technologies. The Carbon Engineering pilot facility in Squamish, Canada represents one of them. Their process is described as follows: An air contractor with a large fan pulls ambient air into the structure, where it passes over a solution of potassium hydroxide. The potassium hydroxide binds to the carbon dioxide molecules and traps them in a liquid solution as carbonate salt. That solution goes through several other chemical processes until the carbon and oxygen atoms become pure compressed carbon dioxide gas to be stored or used to make fuel. The company believes that it can capture carbon dioxide at a cost of $100 per ton. The company is currently working to build industrial-scale DAC facilities that will capture one million tons of carbon dioxide per year, which is equivalent to the work of 40 million trees. The Carbon Engineering study plant was partially funded by the US Department of Energy and by private investors such as Bill Gates.

The Climeworks DAC facility outside Zurich represents the second kind of technology. The facility opened in 2017 and was designed to capture carbon dioxide directly from the atmosphere for sale to third parties. A spokesperson for Climeworks described the process on the website as follows: The ambient air is pulled into the system, passing over a saturated filter, it is then heated to about 100 °C. The carbon dioxide is then released from the filter and collected as concentrated carbon dioxide gas to be supplied to customers or for negative emissions technologies. The remaining air, now free of carbon dioxide, is then released back into the atmosphere. This continuous cycle is then ready to start again. The filter is reused and lasts for several thousand cycles.

This system was expected to capture 800 to 900 tons of carbon dioxide per year at a cost of $500 to $600 per ton. The system is currently transporting the captured carbon dioxide via underground pipes to a greenhouse operated by Gebrüder Meier Primanatura to be absorbed by growing vegetables. The DAC plant sits on top of a waste heat recovery facility that powers the entire process.[109] In April 2020, Climeworks announced on its website that it is working with Karlsruhe Institute of Technology in Germany to use the carbon dioxide captured by their equipment to create carbon black, a vital resource used in electronics and other industrial processes. In 2019, the company began operating a second DAC plant in Hellisheidi, Iceland, capturing 50 metric tons of carbon dioxide per year. The captured gas in this installation is injected underground along with water, where it reacts with basalt and turns into rock. Their third plant is located in Italy, where it is designed to capture 150 metric tons of carbon dioxide, which is to be converted to methane that can be used to power trucks. Below is a photo of what a DAC plant might look like.

DIRECT CARBON DIOXIDE CAPTURE

From Carbon Engineering.com website

These technologies are highly energy-dependent and very costly to operate. Without strong market demand for the byproducts from the captured carbon dioxide or significant subsidies from the international community, this technology will not expand as quickly as needed. The International Energy Agency estimates that we need to capture 350 million metric tons of carbon dioxide per year by 2030, either by CSS or DAC. The international community captured 35 million metric tons in 2019, which is no more than a drop in the bucket. The IPCC has made it clear that we cannot reach our 1.5°C Paris Agreement goal without *negative carbon emissions.* This means all the carbon-mitigation efficiencies we might deploy between now and 2030 will not get us to our goal. We must actually *remove* atmospheric carbon dioxide to avoid the unthinkable effects of additional planetary warming. See the graphic from the World Resources Institute below. Whether DAC technology is capable of being scaled to bring to market at reasonable pricing and with reasonable energy consumption remains to be determined.

Staying Below 2 Degrees of Global Warming

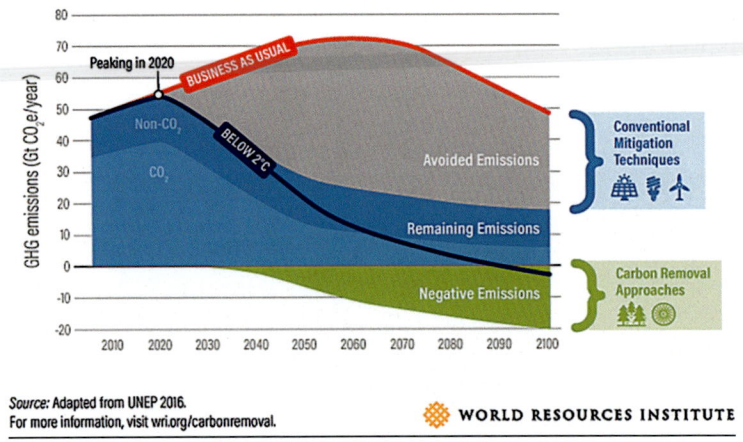

From World Resources Institute, as adapted from United Nations Environment Programme, by Creative Commons license

SOLAR GEOENGINEERING: Solar geoengineering is also called climate engineering or solar radiation management (SRM). This technology primarily involves large-scale efforts to address planetary warming by reducing the amount of solar radiation striking the face of our planet or by reflecting solar radiation into the atmosphere. Most of the work done in this area is hypothetical, based on computer modeling, and very little fieldwork has been accomplished. The basic ideas are as follows: (i) releasing aerosols in the stratosphere to emulate cooling from a large volcano; (ii) marine-cloud-brightening by ships at sea spraying salt water at clouds to make them larger, brighter, and better able to reflect away radiant energy; (iii) maximizing high albedo (very reflective) crops and buildings; (iv) ocean mirroring accomplished by ships churning up microbubbles to make white sea foam which can reflect away solar energy; (v) cloud-thinning by seeding cirrus clouds with dust to make them disappear (cirrus clouds have a net warming impact); and (vi) creating space shades by placing giant mirrors in orbit, which can reflect away solar energy before it even hits the surface of Earth. The graphic below depicts each of these technologies.

Other geoengineering ideas being explored, but not shown in the graphic, include the following: (i) firing sulfate aerosols into the stratosphere to create cloud-like structures that would reflect sunlight into space, (ii) burying large quantities of charcoal under the ground of grazing fields to increase the capacity of the soil to act as a carbon sink, (iii) depositing large quantities of limestone in oceans to increase the alkalinity of the water, thereby improving the ability of our oceans to act as a carbon sink, and (iv) fertilizing the oceans with iron to promote the growth of algae which has a high capacity to sequester carbon.

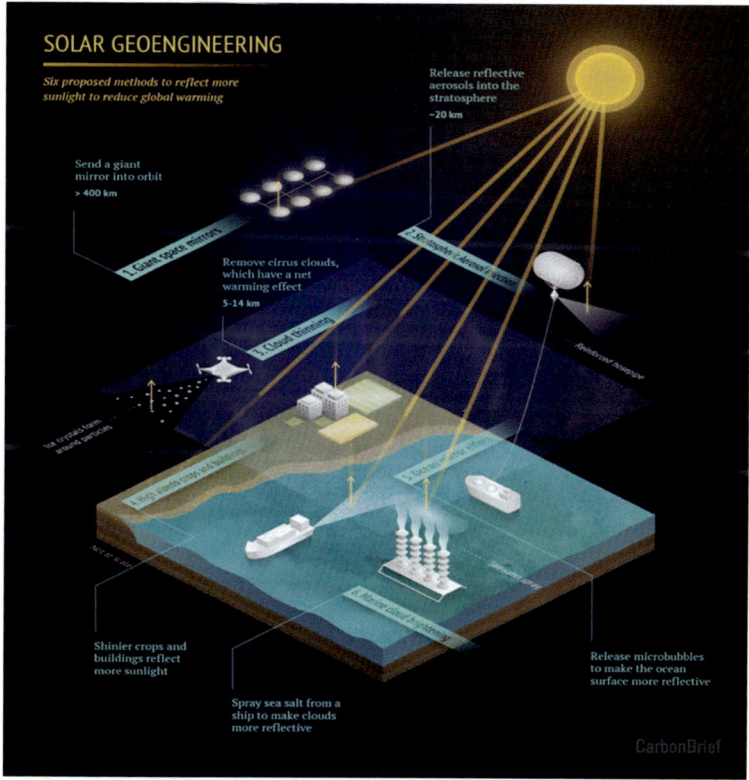

Solar Geoengineering options. Graphic by **Rosamund Pearce** for Carbon Brief.

One significant obstacle to geoengineering technology has been the legitimate fear of unintended consequences from these proposed technologies. Another is the challenge of governance—who is responsible for regulating and implementing this technology? As with multinational surface water flows, geopolitical issues have to be addressed before we can implement any of these technologies. In 2019, Switzerland and ten other countries submitted a proposal to the United Nations requesting that the U.N. Environment Programme (UNEP) prepare a comprehensive assessment of various geoengineering methods. The report is due in August 2020 and is expected to cover the underlying science and technology, as well as ideas for governing and deploying this technology on a World-wide basis.

THE ULTIMATE SWISS ARMY KNIFE: Two of the projects that excite me are multitasking systems developed by Dr Jonathan Trent called OMEGA and UpCycle Systems.

OMEGA stands for Offshore Membrane Enclosures for Growing Algae. OMEGA is a system for producing biofuels using municipal wastewater pumped through floating photobioreactors containing algae. This floating system is powered by solar collectors, and wave action from the water. This project began at NASA in 2008 as a biofuel project under the direction of Dr. Trent. The system places fast-growing freshwater algae in plastic photobioreactors located on floating docks anchored in a protected bay. Wastewater sludge from local wastewater treatment facilities provide the nutrients, water and carbon dioxide for algae growth. The saltwater in the bay controls the temperature inside the photobioreactors and if any algae escapes from a damaged photobioreactor, it will die in the saltwater. The algae purify the wastewater of harmful chemicals in the process of taking up its nutrients. In the rendering on the following page, you can see the bioreactors attached to the floating system.[110]

The OMEGA project received funding for feasibility studies from NASA and the California Energy Commission. It has morphed into OGI, Omega Global Initiative, with plans to scale this project to serve coastal communities. Quoting Dr. Trent:

From the broadest perspective, it seems to me we're standing on a threshold now that is arguably one of the most important in the history of civilization, comparable to the transition from hunting and gathering our food to cultivating it. We now need to make that same transition for energy. We can no longer hunt and gather it, we need to cultivate it, and we need to cultivate it in sustainable and environmentally conscious ways. If we can find the pathway to this transition—and we don't have much time to do it—it will be our legacy for future generations. If we do not at least try, then what?[111]

OMEGA pilot system showing bioreactors. From US-NASA as open source materials

Dr. Trent is now working on a land-based system that integrates livestock agriculture with aquaculture and some of the elements from the OMEGA system. The system proposes to create a connection among animal manure, energy, animal feed, and clean water to improve the efficiency and sustainability of animal farming. In this system, the manure from livestock is collected and used to make a carbon- and nutrient-rich sludge. The sludge is clarified via an electrocoagulation treatment that removes suspended solids, heavy metals, emulsified oils, bacteria and other contaminants so it can be used to cultivate microalgae in photobioreactors floating on a pond nearby. The remaining sludge is then used to produce biogas and carbon dioxide in an anaerobic digester. The biogas is used to produce heat and electricity for the system and the carbon dioxide being transferred to the photobioreactors where the microalgae is growing. The microalgae are used as feed for the livestock and for the fish being farmed in the pond. The fish wastewater is processed by mechanical filtration and provide water for the livestock and the residue from this filtration process is used in the anaerobic digester to produce biogas. Solar panels and wind turbines provide the heat and electricity.[112]

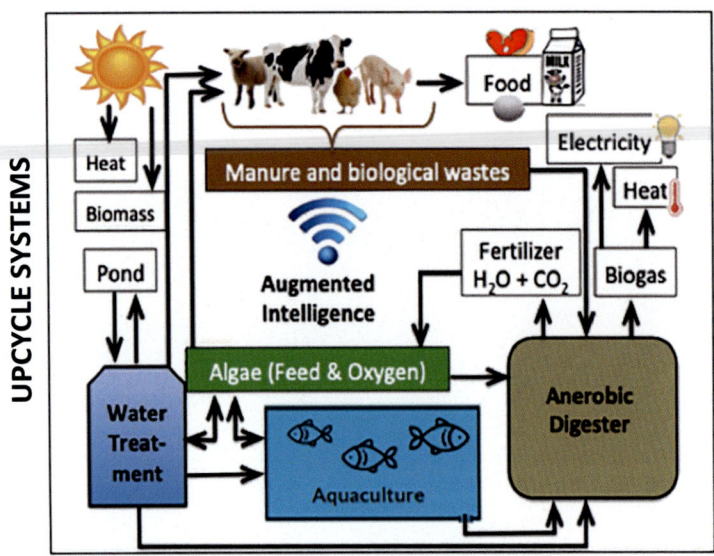

UpCycle Systems: M-Powered Farm Project Methods. Published by National Aeronautics and Space Administration as open source materials

Another quote from Dr. Trent:

We must secure our food, water, and energy systems by upcycling. The UpCycle Systems (UCS) solution: UCS is a food, water, and energy system that integrates livestock farming with aquaculture and algae for food, combined with high-tech water treatment for treatment and reuse, anaerobic digesters for waste-to-energy, and organic fertilizer production. By co-locating and integrating these activities, we take advantage of wastes and byproducts from each subsystem, to improve the efficiency of the overall system and minimize its environmental impact. For years, people have been perfecting each element and combinations of some of the elements of this "ecosystem" of technologies, and now it's time to bring them all together into what we call the "UCS symbiosis."[113]

SPACE-BASED SOLAR POWER (SBSP): We continue to struggle with the fact that we cannot collect solar energy at night or when it is cloudy. Only a small fraction of the sun's solar energy actually strikes Earth's surface. As far back as the 1970s, scientists were exploring the notion of placing the solar collectors in space on an orbit that would provide the collectors with almost continuous exposure to the sun. Since then, many creative ideas of how this might be done have been developed, but they all seem to involve the launching and robotic assembly of several hundred modules to operate in an orbit that tracks the Earth's rotation. Once the system is assembled, the reflectors or mirrors are spread over a considerable area of space, so they can direct large amounts of solar radiation onto the solar panels. The solar panels are designed to convert the collected solar radiation into either microwaves or lasers, both of which are also forms of electromagnetic radiation.

The deep-space systems use microwaves, and the near-space systems use lasers. The converted solar radiation is beamed to Earth, where power-receiving stations collect the laser or microwave signals, convert them to electricity, and feed the electricity to the electric grid. According to the U.S. Department of Energy (US-DOE), the deep space microwave-transmitting satellites would orbit Earth about 22,000 miles above Earth's surface. These microwave transmitting satellites would be very large, with solar reflectors spanning up to 2 miles and weighing almost 90,000 tons. They would be capable of generating enough gigawatts to power a major U.S. city. The US-DOE estimates the cost of launching, assembling, and operating a satellite such as this to be in the tens of billions of dollars, and it could require as many as 40 launches to get all necessary materials in space. On Earth, the antenna used for collecting the microwave beam would be anywhere between 2 to 6 miles in diameter, requiring a significant area of land.

The US-DOE describes the near-space laser transmitting satellites as operating in a low orbit about 250 miles above the Earth's surface, with a weight of fewer than 11 tons. This design is estimated to cost in the range of $500 million to launch and operate. It would be possible to launch the entire self-assembling satellite in a single rocket, drastically reducing the cost and time to production. Also, by using a laser transmitter, the beam will only be about 6 feet in diameter, instead of several miles, a significant reduction. One of the shortcomings of this lighter, less costly system is the lower per-unit capacity, which would require a fleet of satellites to be launched and operated together.[114]

The biggest hurdle has been the cost to fund such an ambitious project, but currently, Japan, China, the United States, and Russia are pursuing the development of this technology. In 2019, China announced that it is planning to build the world's first solar power station to be positioned in Earth's orbit. California Institute of Technology is working to "conceive, design, and demonstrate a scalable vision for a constellation of ultralight modular spacecraft that collect sunlight, transform it into electrical power, and wirelessly beam that electricity to Earth. The basic module of this future solar power system is a giant coilable structure that elastically deploys after launch into orbit and is made of paper-thin materials of high stiffness."[115]

This is an exciting technology, but actually not at all new. The idea first appeared in 1968, in a technical article written by US aerospace engineer Peter Glaser entitled "Power From The Sun: Its Future" published in the journal *Science*. I have read that the idea of collecting solar energy in space and wirelessly transmitting it to Earth was first described by Isaac Asimov in 1941 in a short story entitled "Reason." In thinking about this technology, one comes to realize that much of the component technology for these systems already exists, such as photovoltaic cells, satellite technology, and wireless power transmission.

GOVERNMENT FUNDING: Before leaving this chapter, I have to mention government subsidies again. I have a difficult time watching massive amounts of taxpayer dollars directed to undesirable activities. Presumably, government subsidies are provided to encourage the subsidized behavior. Subsidies are any benefit that reduces the cost of doing business, such as direct grants, loan guarantees, and protective tariffs. In 2009, all of the G7 countries pledged to reduce subsidies to companies in the fossil fuel industry. The International Monetary Fund has been tracking fossil fuel subsidies, and in its most recent working paper dated May 2019, it reported as follows: "Globally, subsidies remained large at $4.7 trillion (6.3% of global GDP) in 2015 and are projected at $5.2 trillion (6.5% of global GDP) in 2017. The largest subsidizers in 2015 were China ($1.4 trillion), United States ($649 billion), Russia ($551 billion), European Union ($289 billion), and India ($209 billion). *Efficient fossil fuel pricing in 2015 would have lowered global carbon emissions by 28% and fossil fuel air pollution deaths by 46%, and increased government revenue by 3.8% of GDP* [emphasis added]."[116]

In January 2020, several former Federal Reserve chairmen, a group of Nobel Prize winners and previous leaders of the president's Council of Economic Advisers jointly issued a statement asserting that a gradually rising carbon tax would be the most effective lever to reduce carbon emissions at the rate and in the amounts required. When I read about subsidies on the scale of those directed to the fossil fuel industry, all I can think about is how those funds would be better utilized for the common good if used to support the creative and bright minds that are working hard to develop solutions that will help our planet mitigate the accumulated damage and depletion we have collectively spawned.

BOTTOM LINE: TECHNOLOGY: Technology has profoundly changed almost every aspect of our lives, and it seems that the newest technologies disrupt the merely new technologies at warp speed. Consequently, it would not appear to be outside the realm of reason to conclude that technology has the potential to extend the carrying capacity of our planet. But is that sound reasoning? The next chapter will provide some of the answers to that question.

Chapter 7

A NEW ECONOMIC PARADIGM

Let's tie this science and technology together to examine the connection between our established economic systems and our environment. As a global society, we have been on a journey of unprecedented economic growth, although far too many have not had the opportunity to join us. Many enjoy a standard of living unimaginable as recently as a few generations ago. Conventional wisdom tells us that more economic growth is the answer to poverty, unemployment, crime, income inequality, and every other socioeconomic ill. The thinking is that if we continue making the pie bigger, everyone will benefit by getting a larger piece. Consequently, countries strive for the growth of their respective gross domestic product (GDP) year after year and boast their economic growth as an indicator of national achievement in multiple realms of endeavor. But does GDP growth really increase the well-being of a population? GDP is essentially the aggregate market value of all products and services bought and sold during a defined period by private individuals, companies, and government entities. GDP grows when there is an increase in the production and consumption of goods and services. Measuring our success by *growth* of GDP requires companies to produce more each year than the prior year. To ensure that this increased production of goods and services gets consumed, companies must find new markets and new consumers by way of marketing and advertising.

Political and corporate leaders have begun using the terms "sustainable growth." This term is intended to denote economic growth that does not put pressure on the environment by way of resource depletion and waste production. I have mentioned throughout this book that the biosphere is the source, the sole source, of the energy and materials required to produce all the goods and services required to sustain life. The notion of "sustainable development" is based upon the assumption that economic growth can take place indefinitely if we just apply technology that will help us become more efficient in our use of energy and resources.

The notion of "sustainable growth," as used above, fails to take into account that every economic system is essentially a throughput system. It utilizes energy and ecological resources to produce goods and services. At the same time, it generates waste and pollution from the production, distribution, and final consumption of goods and services. It is a simple formulation: *energy + raw materials in-->goods, services, and waste out.* As an economy grows, throughput has to increase as more materials and energy go in, and more products, services, and waste come out. On the input side, energy may become more efficient, and technology may develop ways for us to deplete our raw materials at a slower rate, but the output will always include increasing amounts of goods, services, and waste. Additionally, technology that produces efficiency in production will buy us time, but the end result will be the same—we must scale our economic activities to avoid outstripping our ecological resources and services.

The fundamental question that must be answered boils down to whether we can have continuous economic growth *and* environmental sustainability. If you answer that question in the negative because continuously increasing throughput will ultimately diminish our natural resources and disrupt our natural cycles, then the next question is whether world populations can enjoy prosperity and well-being *without* economic growth. I think by the end of this chapter, you will answer that question in the affirmative. You will conclude, as I have, that the term, "sustainable growth" might be the worst oxymoron of our time and that GDP should be benched for the rest of the game, or at least demoted to play second string.

DR. DALY'S STEADY STATE ECONOMY: Dr. Herman Daly, the pioneering ecologist and economist whose notion of sustainability we have been using throughout this book, has been turning the GDP model on its head for decades. As early as the 1960s, Dr. Daly saw that the economy is a subsystem *contained within and sustained by* the larger biosphere. He also saw that because our biosphere is made up of finite raw materials and balanced ecosystems, continuous economic expansion will only convert more and more of the materials of those ecosystems into products and services for humans. He viewed this expansion to be disruptive to our ecosystems and fundamentally constrained by the biophysical limits of our biosphere. He saw how economic expansion, as measured by GDP growth, would make it increasingly challenging to achieve an equitable distribution of wealth between generations, with raw materials being consumed too quickly to regenerate for use by the next generation. He also saw that GDP growth did little to promote an equitable distribution of wealth, income, healthcare services, and education among members of the current generation. The following quote from Daly summarizes his view. "We convert too much of nature into ourselves and our stuff, and there's not enough left to provide the biogeochemical life-support services that we need. Standard economics does not have any mechanism to register the cost of the economy's scale, relative to the biosphere. Prices don't do that. They just measure the scarcity of one resource in relation to another, not the scarcity of all resources relative to the economy's total demand."[117]

The graphic on the following page is a simplified explanation of Dr. Daly's elegant economic model. The biosphere provides the input to the economy; this would include the energy, minerals, metals, water, and other raw materials (economists call this "natural capital") required for the production of goods and services. In turn, the economy produces goods and services for consumption by us, along with waste and pollution generated during production and distribution. In Dr. Daly's universe, many goods would be produced with the end purpose of having their components returned to the production cycle as raw materials, but his concept is far more than a circular economy.

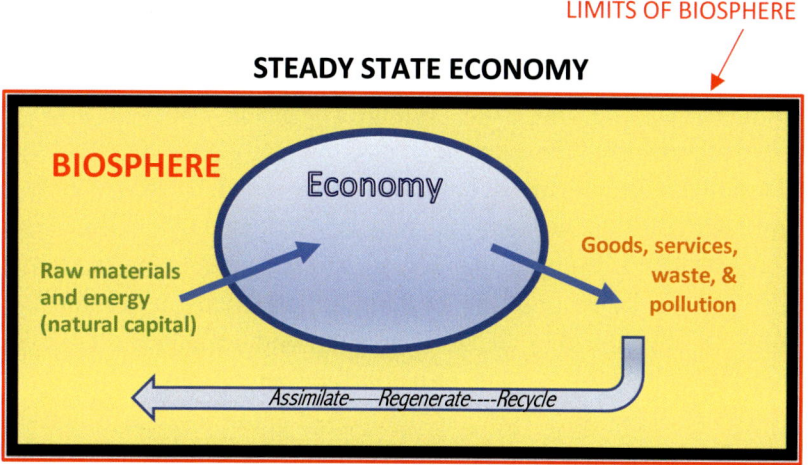

Modified rendition of chart depicting steady state economy by H. E. Daly

The crux of Dr. Daly's argument is that even with improved efficiency, increased production and consumption will ultimately expand the economy up to and beyond the limits of the biosphere. Imagine the blue oval above reaching up to and beyond the four sides of the black and red rectangles. With continuous economic expansion, we will, at some point, find our ecological systems and resources so disrupted that the biosphere can no longer provide its critical life support functions. Only if countries maintain economic activity at levels that fluctuate within a range that is capable of consuming natural capital sustainably (see Chapter 1 definition by Dr. Daly), can we expect to achieve long-term prosperity. For Daly, the economic goals of the developed world need to shift from maximizing to optimizing, from a quantitative growth model to a qualitative regenerative model. *A qualitative regenerative model produces global economic prosperity and social gains without increasing the input of natural capital.* Part of this transition will require that we modify the metrics by which we measure our economic progress to account for and measure these qualitative components. If the purpose of an economy is to sustain human well-being, then it follows that the economy must account for *all* of the components that create and sustain human well-being, not just our financial transactions.

LIMITATIONS OF GDP: As mentioned earlier, the usual gauge of a country's progress is the growth of its gross domestic product (GDP). It is a measure of all economic transactions, good, bad, and ugly. This metric was developed during the Great Depression early in the 20th century to provide lawmakers with a tool to inform their policy decisions. GDP *excludes* the value of everything that is outside the scope of monetized transactions. It also indirectly discounts the value of beneficial components of an economy that generally are not greatly profitable, such as hospitals and schools. GDP also ignores the costs of environmental impacts, such as resource depletion and pollution, and other social costs of our economic transactions. It is akin to a business only tracking its gross income and making its business decisions exclusively on that basis, rather than tracking and studying each expenditure contributing to the success of the business. The GDP model tacitly assumes infinite growth of the economy and ignores the fact that much of what we consider to be economic growth is actually a loss (e.g., resource extractive activities). The following quote from the World Bank explains the concept well.

> Countries rely on GDP as a measure of its economic performance. However, GDP only measures current income and production. It tells us nothing about income for the long term. It does not answer questions like: are income and growth sustainable? Will the same level of income be available for our children? GDP says nothing about the assets that underpin this generation of income. For example, when a country exploits its minerals, it is actually diminishing wealth. The other major limitation of GDP is its poor representation of natural capital. In addition, GDP does not account for contributions to the well-being of the environment provided by forests, wetlands, and agricultural land. Natural capital includes the resources that we easily recognize and measure, such as minerals, agricultural land, and fisheries. It also includes ecosystems producing services that are often 'invisible' to most people such as air and water filtration, flood protection, and habitat for fisheries and wildlife. We often take these services for granted and do not know what it would cost if we lose them. Natural capital accounting can provide detailed statistics to better manage these natural

resources and ultimately ensure sustainable growth of the economy.[118]

Our stubborn adherence to GDP has not served us well as a society, as shown by the following examples.

- First, research by Robert Costanza and Ida Kubiszewski has demonstrated that continued economic growth does not necessarily equate to continued personal well-being. Dr. Costanza is a highly regarded ecological economist, author, and a co-founder of the International Society for Ecological Economics. Dr. Kubiszewski is an environmental scientist who has written numerous scientific articles and books on the subject. Both are currently professors at Crawford School of Public Policy at the Australian National University. Their research has shown that in country after country, the two measures, per capita economic growth and reported well-being, will trend upward in lockstep, but only until a certain level of per capita GDP is reached. At that point, the two measures split, with per capita GDP continuing on its upward trajectory and reported well-being flattening. It appears that humans require a certain level of income to gain a sense of well-being, but once that level is attained, more income does not produce additional reported well-being. [119] For instance, in the United States, per capita GDP doubled between the years 1950 to 1975, but reported well-being remained flat during that same time.

- Second, it is generally accepted economic theory that without sustained economic growth, there will be widespread poverty and misery. Interestingly, between 1970 and 2017, the United States total GDP increased from $1 trillion to $19.5 trillion, while the US Census Bureau shows that poverty rates during that same time remained flat at about 12%. Additionally, during that same period, the United States had the second highest mental health and substance abuse rates in the world, remaining generally flat from 1970 to 2017 (approximately 17.3% according to the Institute for Health Metrics and Evaluation in Seattle, Washington).

• Third, continuous economic growth requires continuously increasing energy use and has become the primary contributor to our ever-worsening climate disruption. While some say that we must choose between economic growth and mitigation of greenhouse gas emissions, we should collectively acknowledge that we are going to experience a reduction in economic growth one way or another. It may cost us today to expend funds to reduce and mitigate emissions, but it is and will continue to cost us not to do so. One of the more conservative estimates is that by the end of this century, more than 10% of our GDP will be spent on repairing the adverse impacts of climate system disruption (e.g., storm damage, reduced agricultural productivity, and rising sea levels), while the cost of mitigating the causes of climate disruption today has been estimated at 1.5 to 2% of GDP.

• Fourth, while GDP in the United States has increased astronomically, our overall well-being as a population has not. Compared to other affluent, developed countries, the US currently has the highest healthcare costs, one of the highest chronic disease and obesity burdens, one of the lowest life expectancy, the highest suicide rates, very high income inequality, and one of the highest incarceration rates.

• Fifth, because GDP ignores the value of almost all ecosystem services, governments that rely solely on GDP for policy decisions will be without the full gamut of tools available to develop public policy directed at improving the well-being of its population and the planet. We will discuss some of these tools in this chapter.

• Even Simon Kuznets, the chief architect of GDP, advised against use of GDP as a measure of economic welfare.

ACCOUNTING FOR THE COMMONS--NATURAL CAPITAL ACCOUNTING: Another significant shortcoming in traditional GDP accounting is that it does not account for the *commons,* which is an old English legal term referring to collective ecological resources shared by many, usually with minimal restriction. In an article entitled "The

Tragedy of the Commons," published in *Science* in 1968, ecologist Garrett Hardin described the problems associated with shared ecological resources. He compared shared resources to a common grazing pasture. In his analogy, everyone with rights to the pasture grazes as many animals as possible, acting in self-interest for the greatest personal gain. Eventually, they collectively use up all the grass in the pasture, and the shared resource is depleted. He believed that shared environmental resources tend to be overused and ultimately depleted, with each party motivated by private self-interest in deciding how much of a resource to use. Today, the costs of these individual decisions are borne by all of us. Examples include the collapse of the codfish population resulting from overfishing in the Grand Banks or the loss of carbon sinks from uncontrolled logging in tropical rainforests. The time has come to account for the commons.

In 1997, Robert Costanza and his research associates attempted to quantify the value of the global "commons," the total of all global ecosystem services, a rather formidable task if you think about it.

> The services of ecological systems and the natural capital stocks that produce them are critical to the functioning of the Earth's life-support system. They contribute to human welfare, both directly and indirectly, and therefore represent part of the total economic value of the planet. We have estimated the current economic value of 17 ecosystem services for 16 biomes, based on published studies and a few original calculations. For the entire biosphere, the value (most of which is outside the market) is estimated to be in the range of US $16 to $54 *trillion* per year, with an average of US $33 *trillion* per year. Because of the nature of the uncertainties, this must be considered a minimum estimate.[120]

Dr. Costanza advises us in his article that the biosphere valuation was about twice the global GDP valuation for that year. You should be able to name most of the 17 ecosystem services Costanza used in his study. They include soil formation, nutrient recycling, storage and purification of fresh water, raw materials, waste detoxification, erosion control, and greenhouse gas regulation. This valuation was updated by Dr. Costanza and his research associates in 2014 to US

$125 to $145 *trillion* per year. While this is recognized as not a perfectly accurate accounting, it has been used by various countries in their well-being accounting, land use planning, true cost accounting, and policy decisions.

As another example of the value of accounting for the commons, the UN-FAO has estimated the hidden annual environmental cost of global food lost or wasted at $2.6 *trillion*. This analysis is well-documented in a detailed report published by the UN-FAO entitled *Food Wastage Footprint Full-Cost Accounting.* [121] The FAO report recognized that a full one-third of all food produced is wasted or otherwise lost. The market value of the food lost or wasted was assessed at $1 *trillion*, but with true-cost accounting, we learn that this $1 *trillion* loss is, in fact, a $2.6 *trillion* loss. Much of the difference in the two valuations was in the cost of the environmental damage attributable to the lost or wasted food. The following are some of the factors that the FAO examined and quantified: greenhouse gas emissions, water loss, soil erosion, and nitrogen and phosphorus pollution attributable to the lost and wasted food.

ALTERNATIVE MEASURES OF ECONOMIC PROGRESS: If we wish to craft public policy that will better protect our ecological systems and improve the quality of life for current and future generations, we must develop and implement new metrics to track both market and non-market activities that include the elements that we value. The business guru Peter Drucker is attributed with wisely stating that "If you don't measure it, you can't improve it." The *Genuine Progress Indicator* (GPI), described below, provides just such a tool for examining and quantifying these additional elements. These metrics provide policymakers with a means to objectively evaluate and quantify the trade-offs they confront when making policy decisions. GPI should inform us of our progress and setbacks.

This metric is designed to measure and track not only the economic growth of a country or other governmental entity, but also environmental and social elements. Broadly speaking, GPI is calculated by starting with total private and public monetary expenditures (essentially, GDP), and then adjusting that number upwards and downwards by accounting for costs associated with non-monetized products and services that generate benefits or harms. Examples of these adjustments include (i) the monetary value of non-market work (e.g., volunteer work, education), (ii) the monetary costs associated with activities such as commuting or crime (e.g., pollution and energy depletion, medical costs and law enforcement costs), and (iii) the value of services provided by natural capital, such as wetlands, cropland, and forests (e.g., water purification, food production, carbon sinks). Each of these considerations is reflected in monetary terms as a component of the GPI. As an example, GPI would be reduced when crime rates increase or mass transit ridership decreases. On the other hand, GPI would be increased when unemployment rates decrease or atmospheric carbon dioxide levels decrease. Another example would be pollution from an oil spill. While GDP would consider the cost of clean-up as economic *value* to be added to GDP, GPI would count this as a *loss,* measured by the projected cost to clean up and the value of any ecosystem damage.

Several US states and countries have adopted this means of accounting to provide their citizens and decision-makers the tools to make better choices about their shared future. This tool provides well-documented and quantified data regarding key indicators of environmental sustainability and quality of life. GPI is an unbiased, quantified quality-of-life metric that assesses current net income by incorporating adjustments for economic, environmental, and social values and costs. It finally recognizes and accounts for what economists call the "externalities" of economic activity. These are consequences, positive or negative, of commercial activity that affect third parties but are not reflected in the cost of the goods or services.

The first graphic on the following page shows each of the progress indicators identified by the state of Utah. The graphic was prepared by the Utah Population and Environment Council (UPEC) for Utah's GPI assessment. The second graphic, also prepared by UPEC, reflects the valuation assigned to each of the identified progress indicators, to be treated as deductions or additions to GDP. [122] You can see from both these graphics how GPI would be a useful tool to quantify and track over time economic losses associated with activities like water pollution or mineral extraction, as well as economic gains associated with replanting forests or reducing carbon emissions. Each of these activities was quantified by UPEC. As an example, the UPEC reports that Utah's 2007 GDP was reduced by $408 million due to poor air quality and loss of farmland. On the other hand, Utah's 2007 GDP was enhanced by $24.8 billion from the value of ecosystem services provided by Utah's wetlands, forests, and deserts.

GENUINE PROGRESS INDICATORS (GPI): UTAH

+ = addition to GPI (value) − = deduction from GPI (cost)

Berik, G. and E. Gaddis. The Utah Genuine Progress Indicator (GPI), 1990 to 2007
https://utahpopulation.org/wp-content/uploads/2014/11/Utah_GPI__Report_v74_withabstract.pdf

VALUE OF GENERAL PROGRESS INDICATORS (GPI): UTAH

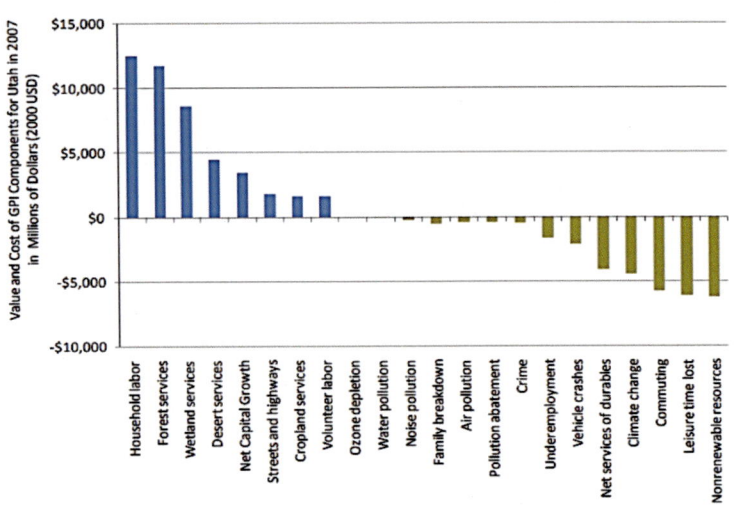

Versions of this accounting system have been formally adopted by several countries and a few US states (e.g., Vermont, Maryland, and Utah). In addition, a large number of states and countries have commissioned GPI studies to examine their respective progress over the past decades. Whether adopted as the accounting method of choice or merely used as a report card for taking stock at a defined point in time, GPI has been found to be a useful tool to inform policy.

TRUE COST ACCOUNTING: It is easy to overlook the rudimentary notion that all economic activity ultimately depends on nature. Stated another way, our ecological resources and services provide us with everything required for our economy to function. The value of the ecological resources and services employed in manufacturing and distributing goods and services have historically not been recognized by companies as an operating expense. Omitting the value of natural capital means that the items we purchase are deeply discounted from their actual cost to produce and that certain companies may not be as profitable as their operating statements would suggest.

Think back to the chapter on food production impacts. Livestock agriculture is responsible for a wide assortment of externalized costs, such as the cost of ground water pollution from manure management systems, damage to soil ecosystems, surface water pollution from the use of excessive fertilizers, greenhouse gas emissions, and loss of birds and honeybees due to overuse of pesticides. Just think what these food items would cost us if these externalized costs were incorporated into the price of the products. Now let's look at this from Farmer Joe's perspective: The externalized costs of Farmer Joe's farming operation can be calculated, and we will call that number X. For example, when Farmer Joe pollutes the river with manure runoff and the town downstream has to remove toxins from the river water, the town's primary water source, you have an externalized cost that can be quantified. Now we will calculate Farmer Joe's share of the $38 billion in government subsidies, which we will call Y. Farmer Joe has already calculated what he considers to be his direct operating costs, such as labor, fertilizer, transportation, and fuel, and we will call these costs Z. Add X, Y, and Z together, and you will have Farmer Joe's true operating costs. Now subtract that amount from Farmer Joe's gross operating income. Do you think Farmer Joe is making a profit? Would you be shocked by the true cost of a steak produced by Farmer Joe if he based his prices on his true expenses? Do you think he might employ more sustainable farming practices if he had to pay the true costs of production? Before you say that you don't want to pay more for your steak, keep in mind that your taxes are already funding the subsidies paid to farmer Joe, along with the cost of cleaning the water supply of the downstream town. In the final analysis, you and I end up paying for most of the externalized costs, although not necessarily at the grocery store. Natural capital, such as soil, clean air, clean and abundant water, is owned by all of us, it is the commons, and when a single individual or company damages those assets, we all suffer a loss. Livestock agriculture is not the only unprofitable and unsustainable business sector being propped up by you and me.

It is becoming increasingly evident that we need a universally accepted way of accounting that takes into consideration the value of the ecological goods and services consumed, as well as the waste and damage incurred by our ecosystems in the production and distribution of our goods and services. Until we have a more accurate way of accounting for our economic activity, how will we know if we are actually improving our economic and environmental conditions or merely depleting resources? GDP will definitely not tell us. The effort to refine this accounting with GPI is well underway, but until it is honed and accepted by the global community, we will not have this vital tool to address our environmental sustainability concerns. The best incentives in today's world are financial ones; however, until the global community agrees on a methodology for monetizing environmental harms and benefits, there will be no effective tool to reward sustainable industrial practices and to discourage unsustainable practices. If we want an economy that can sustain us for generations to come, we must have a measurement system that will account for the things we value and discourage the things that harm us. For policymakers, this void is significant. Remember Drucker's quote, "If you can't measure it, you can't improve it."

PLANETARY BOUNDARIES: In 2009, a group of 26 bold environmental scientists led by Johan Rockström from the Stockholm Resilience Centre and Will Steffen from Australian National University worked together to define the planetary life-support systems essential for survival. They next quantified the safe consumption/degradation limit of each of these systems, and finally, they determined our global standing with respect to the defined safe boundary for each of these systems. The original 2009 systems included the freshwater cycle, the phosphorus cycle, the nitrogen cycle, ozone depletion, ocean acidification, climate change, chemical pollution, atmospheric pollution, biodiversity loss and land-use changes.

In 2015 the group restated the systems a bit differently, but they incorporated all of the original systems. The group saw that there was a critical need to provide policymakers, lawmakers, and the general public a science-based framework that would identify the most critical "biophysical processes that regulate the stability of the Earth system, which if disrupted past a defined point, could disturb Earth's services and resources at the planetary level." [123] The group was able to quantify the safe levels of operating for each ecological service *and* quantify the extent to which human activity has exceeded those levels. They called these limits "planetary boundaries" (PB). "The PB framework aims to help guide human societies away from such a trajectory by defining a 'safe operating space' in which we can continue to develop and thrive. It does this by proposing boundaries for anthropogenic perturbation of critical Earth-system processes. Respecting these boundaries would greatly reduce the risk that anthropogenic activities could inadvertently drive the Earth system to a much less hospitable state." [124] The Planetary Boundary graphic on the following page depicts each of these Earth systems. The authors help us navigate the planetary boundaries graphic as follows:

> The green zone is the safe operating space, the yellow represents the zone of uncertainty (increasing risk), and the red is a high-risk zone. The planetary boundary itself lies at the intersection of the green and yellow zones. The control variables have been normalized for the zone of uncertainty; the center of the figure, therefore, does not represent values of 0 for the control variables. The control variable shown for climate change is atmospheric carbon dioxide concentration. Processes for which global level boundaries cannot yet be quantified are represented by gray wedges; these are

atmospheric aerosol loading, novel entities, and the functional role of biosphere integrity.[125]

PLANETARY BOUNDARIES 2015

Credit: Steffen, Will, Richardson, Katherine, Rockström, Johan, et al.
https://science.sciencemag.org/content/347/6223/1259855.full

The table on the following page, from this same project, quantifies two of the Earth processes: climate change and ocean acidification. It identifies the control variable or proxy they are using to measure the process, the quantified threshold for each process, and the current status of that process. For instance, the level of atmospheric greenhouse gases, expressed as carbon dioxide equivalents, is the control variable for tracking climate change (columns 1 and 2). The safe operating zone for atmospheric carbon dioxide equivalents is 350 parts per million (column 3), and the current concentration in 2015 was 398.5 parts per million (column 4), outside the safe operating zone. A link to the entire article is in this endnote.[126] Dr. Rockström has written a very accessible book on this subject as well, cited in endnote number 3 in Chapter 1

Climate change (R2009: same)	A Atmospheric CO_2 concentration, ppm B Energy imbalance at top-of-atmosphere, W m-2	A 350 ppm CO_2 (350-450 ppm) B. +1.0 W m-2 (+1.0-1.5 W m-2)	A. 398.5 ppm CO_2 2.3 W m-2 (1.1-3.3 W m-2)
Ocean acidification (R2009: same)	Carbonate ion concentration, average global surface ocean saturation state with respect to aragonite (Ωarag)	Sustain ≥80% of the pre-industrial aragonite saturation state of mean surface ocean, including natural diel and seasonal variability (≥80%–≥70%)	~84% of the pre-industrial aragonite saturation state

Credit: Steffen, Will, Richardson, Katherine, Rockström, Johan, et al.
https://science.sciencemag.org/content/347/6223/1259855.full

ECONOMIC GROWTH AND SUSTAINABILITY, HEADED FOR DIVORCE COURT? The planetary boundaries provide us with a tool to determine if economic activity is exceeding the safe operating level required to keep our ecological life support systems in good working order. Each year, businesses strive to produce more goods and services than the year before and to find new markets for their goods and services. If our economic activities cannot keep within our planetary boundaries, then we will continue to diminish and disrupt our natural resources and ecological systems, as demonstrated throughout this book. Many economists and international institutions like the United Nations recognize this, but their response is the notion of "sustainable development." As mentioned earlier, this notion is based upon the common belief that continued economic growth can exist within the limits of our biosphere, provided we increase the efficiency of our production technologies. In response to this notion, Dr. Daly would remind us that it is simply not possible to produce increasingly more goods and services without consuming increasingly more natural capital, with the exhaustion of natural resources and damage to our ecological systems as the only possible outcome. In the years since the notion of green growth was introduced at the 2012 United Nations Rio conference, several research studies have been completed which support Daly's conclusion that separating GDP growth from resource use is not possible on a global scale.[127] Two of these studies are described below.

In 2012, German researcher Monika Dittrich and her team developed a computer model that forecasts future resource consumption. The model showed that if economic growth continues to increase at a rate of 2% to 3% per year from 2012 to 2050, consumption of natural resources will increase from 70 billion metric tons per year to 180 billion metric tons per year over that same period. The team ran the model a second time to see what would happen if every nation on Earth immediately adopted the best industry practices in resource use. The results showed some improvement but were still at twice the sustainable level. For a reference point, global sustainable consumption is estimated at 50 billion metric tons per year.[128]

In 2016, a team of researchers led by Heinz Schandl tested a slightly different premise. The team assumed that all of the world's nations agreed to go beyond existing best practices. They also assumed the imposition of a tax that would increase the global price of carbon from $50 to $236 per metric ton. Their final assumption was that technological innovation would double the efficiency of current resource consumption (in other words, they assumed that technology would reduce resource consumption by 50%). They found that, even with these unlikely and idealized assumptions, if the global economy kept growing by 3% each year, global consumption of natural resources would reach an unacceptable 95 billion metric tons by 2050.[129]

We are on a path that is headed toward bankrupting our biosphere by spending down our natural capital to finance economic growth. The time has come for a new way to view and measure economic progress. It would be comforting to think that we could continue on this same path of GDP growth ad infinitum. As stated earlier, GDP fails to take into account the true cost of producing goods and services by ignoring the value of the natural capital depleted or damaged and other non-monetized factors. The proponents of continuous GDP growth fail to consider that an economy can maintain high levels of employment and productivity without producing more "stuff" for our population to consume. The demand for quarterly profits for shareholders means that many companies (and therefore, many politicians) will be opposed to GPI accounting and true cost accounting. When we consider Dr. Daly's stable-state economy, the merits of true-cost accounting, the genuine progress indicators, and the practical knowledge gained from quantified planetary boundaries, we have to conclude that economic growth cannot be decoupled from ecological resource and system limitations. We have reached the point where the ecological resources used to manufacture the products and services we purchase are worth much more than the products and services. Chew on that for a minute.

BOTTOM LINE: ECONOMICS AND SUSTAINABILITY: The primary takeaway from this chapter is a theme worthy of congratulatory acknowledgment. Today, after decades of innovative thinking by such luminaries as Herman Daly, Robert Costanza, Johan Rockström, Will Steffen and many others, we are able to quantify the biophysical limits of our ecological systems, the economic value of these systems, and the monetary cost of abusing those limits. Paraphrasing Peter Drucker, now that we can measure, we can manage.

Let's conclude this chapter with quotes from Herman Daly and Paul Hawken and a quote attributed to Albert Einstein.

> "The economy is a wholly-owned subsidiary of the environment, not the reverse."—Herman Daly[130]

> "A new type of thinking is essential if mankind is to survive and move toward higher levels."—attributed to Albert Einstein, allegedly in a *New York Times* article dated May 25, 1946

> "We need to revise our economic thinking to give full value to our natural resources. This revised economics will stabilize both the theory and the practice of free-market capitalism. It will provide business and public policy with a powerful new tool for economic development, profitability, and the promotion of the public good."—Paul Hawken[131]

CONCLUSION

You have just completed a study of the ecosystem services that provide life support for our planet. You learned that life on Earth is entirely dependent upon the interworking of various ecological systems and the abundant natural resources with which our planet is endowed. We explored many of these ecological systems and natural resources, developed over millions of years, and studied the many ways they have been disrupted by the activities of humankind. We examined the carrying capacity and biocapacity of our planet, learning that these variables are a function of the size of our population and the level of our consumption. We then examined two specific activities that play a significant role in reducing the carrying capacity and biocapacity of our planet: livestock agriculture and population growth momentum. Interwoven into all of these chapters was the notion that our relentless endeavor to produce economic value has been the primary driver causing injury to our ecosystem services. We then switched gears to take a look at some of the cutting-edge technology that may buy us a bit more time to curb population and economic growth. Finally, we examined economic and natural resource accounting models that can be tools for quantifying, accounting, and managing the environmental impacts caused by economic and population growth.

Humanity has relied upon a relatively stable and predictable climate and a broad diversity of plant and animal life on this planet to attain the high level of civilization we enjoy today. Our trajectory has been one of growth, followed by more growth. You probably noticed that the charts and quantitative references throughout this book, such as population growth, increase in greenhouse gas emissions, and growth of gross domestic product, look very similar. For a long time, they remain moderately flat, but then they spike into an almost vertical line over the most recent 200 years—these are all classic hockey stick graphs. A hockey stick graph depicts exponential growth, where the increases begin to accrue at an accelerating rate. When it comes to your savings, hockey stick growth is brilliant. For example, if you invest $10,000 today at 7% interest, your investment will double every ten years; after 30 years, you will have $80,000 without lifting a finger.

On the other hand, in the case of our ecosystem services, exponential growth has meant that we are borrowing from future generations by consuming renewable resources faster than they can be restored, liquidating non-renewable reserves, and disrupting the ecological cycles that sustain life. In the case of population, exponential growth has created population growth momentum projected to deliver population increases throughout the remainder of this century. Exponential growth on all fronts has undeniably ruled the day, but how long can this growth continue? Is "sustainable growth" actually a thing? It seems that we have reveled in our growth without even a backward glance at the fact that the resources that feed this growth are finite, and the systems that are disrupted by this growth are our life-support mechanisms.

We have seen that persistent economic growth, injurious land-use practices, and overpopulation, are the primary factors that have adversely altered our environment and our life-support mechanisms. Can new technology allow us to continue these practices without adverse effects to our environment? In other words, can technology provide the means for economic growth to decouple from environmental impact? The research discussed in the previous chapter would tell us that decoupling is not possible in a world of finite natural resources and ecological systems. There is a biophysical limit to the amount of use and abuse they can absorb. It would be comforting to consider that technology can eliminate the constraints of the natural world and allow us to go on as we have. On the other hand, technology could slow resource depletion and mitigate environmental damage from certain economic activities. This would buy us more time to scale back population, improve agricultural practices, and bring economic activity into balance with our planet's resources and capabilities. Government financial support would accelerate the research and development of critical technology that could slow our current trajectory. There is no single silver bullet, but a broad assortment of technologies that could be deployed, sooner rather than later. We need to recapture some of the time we have squandered looking for solutions in all the wrong places

There is not a scintilla of scientific evidence for any of us on planet Earth to reasonably expect our planet to accommodate the projected 11+ billion people at levels even approaching the prosperity enjoyed today. We need fewer consumers. Exponential mathematics tells us that we cannot achieve this with current lowered population growth rates. We require *negative* population growth, a fertility rate that is less than replacement rate. Optimally, this would begin in the most affluent industrialized countries, where we can achieve the most significant per capita reduction in ecological footprint.

It is also incumbent upon the affluent industrialized countries to wean themselves from their addictions to economic growth. These countries require a common natural capital accounting system to track and measure economic activity in ways that will account for environmental impacts. Once we can identify where and to what extent we are exceeding the limits of our biosphere, policymakers and lawmakers will have the tools to develop financial incentives and disincentives to guide personal and commercial conduct. Such policies could have the potential to contain our economic activity within a safe operating zone. The road to a healthier environment will be paved with a shared vision of a world where we recognize natural capital as valuable, or in some cases, more valuable, than economic capital and begin to treat it as such in the choices we make at every level of human enterprise.

CITATIONS

[1] Daly, H. E. (1990). Toward some Operational Principles of Sustainable Development. *Ecological Economics*, Vol. 2(1), Pgs. 1–6.

[2] Vandermaesen, T. and Wackernagel, M. (2019). *Living Beyond Nature's Limits*. World Wide Fund for Nature. Retrieved from https://www. footprintnetwork.org/content/uploads /2019/ 05/ WWF-GFN-EU-Overshoot-Day-report.pdf

[3] Rockströmand, J. and Klum, M. (2015). *Big World Small Planet: Abundance within Planetary Boundaries*. New Haven, CT Yale University Press.

[4] Wackernagel, M. and Beyers, B. (2019) *Ecological Footprint: Managing our Biocapacity Budget*. British Columbia, Canada: New Society Publishers.

[5] Malthus, T. (1798). *An Essay on the Principle of Population*. p. 44. Retrieved from http://www.esp.org /books/malthus/population /malthus.pdf

[6] Ibid., p. 68.

[7] Shankman, S. (2019). What is Nitrous Oxide and Why is it a Climate Threat. *Inside Climate News.* Retrieved from https:// insideclimatenews.org /news/ 11092019/ nitrous-oxide

[8] US Environmental Protection Agency. Science Advisory Board. (2011). *Reactive Nitrogen in the United States: An Analysis of Inputs, Flows, Consequences and Management Option*s. (Report no. EPA-SAB-11-013). Retrieved from https://yosemite.epa.gov/ Final%20INC%20Report_8_19_11 (without%20 signatures)

[9] Smit, A.L., Bindraban, P.S., et al. (2009*). Phosphorus in Agriculture, Global Resources, Trends and Developments*. Report to the Steering Committee Technology Assessment of the Ministry of Agriculture, Nature and Food Quality. Retrieved from https://library.wur.nl/ WebQuery/ wurpubs/fulltext/12571

[10] de Boer, M. and Faradji C. (2016*).* How the Great Phosphorus Shortage Could Leave Us all Hungry. *The Conversation*. Retrieved from https:// www.sbs.com.au/ topics/ science/ earth /article2016/ 02/19/how-great-phosphorusshortage-could-leave-us-all-hungry

[11] USGS. Retrieved from https://www.usgs.gov/special-topic/water-science-school/science/where-earths-water?qt-

[12] World Economic Forum. (2019). *The Global Risks Report 14th Edition.* Retrieved from https://www3.weforum.org/docs/ WEF_Global_Risks_ Report_2019.pdf

[13] United Nations Water Agency. (no date). *Water Scarcity.* Retrieved from https://www.unwater.org/ water-facts/scarcity/

[14] Gleeson, T., Wada, Y., et al. (2012). Water Balance of Global Aquifers Revealed by Groundwater Footprint. *Nature,* 448, 197-200,

[15] American Geophysical Union. (2010) *Groundwater depletion rate accelerating worldwide.* Retrieved from https://www.sciencedaily.com/ releases/2010/ 09/100923142503.htm

[16] United Nations Water Agency. Retrieved from https://www.unwater.org/ water-facts/transboundary-waters/

[17] Dental, M. (2018). *Water Pollution: Everything You Need to Know.* National Resource Defense Council. Retrieved from https:// www.nrdc.org/stories/water-pollution-everything-you-need-know#cause.

[18] United Nations Water Agency. Retrieved from https://www.unwater.org/water-facts/quality-and-wastewater/

[19] United Nations. (2010, October 21). Unprecedented Pressures on Farmland—with 30 million Hectares Lost Annually—Poses 'Direct Threat to the Right to Food of Rural Populations', Third Committee Told. [Press release]. Retrieved from https://un.org/press/en/2010/ gashc3985.doc.htm

[20] Pinentel, D. and Burgess, M. (2013). Soil Erosion Threatens Food Production." *Agriculture*, 3(3), 443-463.

[21] Hsu, J. (2019). Don't Panic about Rare Earth Elements. *Scientific American*. Retrieved from https://www.scientificamerican.com/article/dont-panicabout-rare-earth-elements/

[22] Sverdrup, H. and Koca, D., et al., (2014). Investigating the Sustainability of the Global Silver Supply, Reserves, Stocks in Society and Market Price Using Different Approaches. *Resources, Conservation and Recycling,* 83, 121-140.

23 World Economic Forum. (2019). *The Global Risks Report 14th Edition.* Retrieved from https://www3.weforum.org/ docs/WEF_Global RisksReport2019.pdf

24 Carrington, D. (2018). What is Biodiversity and Why Does it Matter to Us? *Guardian.* Retrieved from https://www.theguardian.com/news/ 2018/mar/12/what-is-biodiversity-and-why-does-it-matter-to-us

25 National Center for Atmospheric Research with modifications by University Corporation for Atmospheric Research. (2007) *The Carbon Cycle.* Retrieved from http://scied.ucar.edu/ carbon-cycle

26 Higgins, P. and Harte, J. (2012). Carbon Cycle Uncertainty Increases Climate Change Risks and Mitigation Challenges. *Journal of Climate,* 25, 7660-7668. Retrieved from https://doi.org/ 10.1175/JCLI-D-12-00089.1

27 Hubau, W., Lewis, S.L., et al. (2020). Asynchronous Carbon Sink Saturation in African and Amazonian Tropical Forests. *Nature.* 579, 80-87.

28 LeQuéré, C., Jain, A., et al. (2018). The Global Carbon Budget 2017. *Earth System Science Data Discussions,* 10, 405-448. Retrieved from https://www.earth-syst-sci-data.net/10/405/2018/essd-10-405-2018.pdf

29 Peters, G. and Korsbakken, R. (2018). Global Carbon Dioxide Emissions Rise Again in 2018. *Center for International Climate Research,* Retrieved from https://cicero.oslo.no/en/. Peters, G., Andrew, R., et al. (2019) Carbon Dioxide Emissions Continue to Grow amidst Slowing Emerging Climate Policies. *Nature Climate Change, 10, 3-6.*

30 Masson-Delmotte, V., Zhai, P., et al. (2018) Global Warming of 1.5°C: An IPCC Special Report on the Impacts of Global Warming of 1.5°C Above Pre-Industrial Levels. *IPCC 2018 Summary for Policymakers.* Retrieved from https://report.ipcc.ch /sr15/ pdf/ sr15_ spm_ final.pdf and Akerman, F., Stanton, E., et al. (2008). The Cost of Climate Change. National Resources Defense Council. Retrieved from https:// www.nrdc.org/ sites/default/files/ cost.pdf

31 Myers, N. (2005). "Environmental Refugees: An Emergent Security Issue." Retrieved from https://www.osce.org/eea/14851?download=true

32 Tilic, E., Goffreidi, S. et al. (2020). Methanotrophic bacterial symbionts fuel dense populations of deep-sea feather duster worms

(Sabellida, Annelida) and extend the spatial influence of methane seepage. *Science Advances,* 6 (14), 8562. Retrieved from https:// advances.sciencemag.org /content/6/14 /eaay8562.abstract

[33] Alvarez, R., Zavala-Araiza, D., et al. (2018). An Assessment of Methane Emissions from the US Oil and Gas Supply Chain. *Science,* 361(6398), 186-188.

[34] Hmie, B., Petrenko,V.V, et al. (2020). Preindustrial CH4 Indicates Greater Anthropogenic Fossil CH4 Emissions. *Nature, 578, 409-412.*

[35] Pearce, F. (2016). What is Causing the Recent Rise in Methane Emissions? *Yale Environment 360* Retrieved from https:// e360.yale.edu/ features/ methaneriddlewhatiscausingtherisein emissions.Also, Nisbet, E.G. et al., (2016), Rising Atmospheric Methane: 2007-2014 Growth and Isotopic Shift. *Global Biogeochemical Cycles*, 30, 1356-1370. Retrieved from https:// doi.org.10.1002/2016GB0005406

[36] US Environmental Protection Agency. *Overview of Greenhouse Gasses, Nitrous Oxide Emissions.* Retrieved from https://www.epa.gov/ ghgemissions/overview-greenhouse-gases

[37] Millar, N., Doll, J., et al. (2014). Management of Nitrogen Fertilizer to Reduce Nitrous Oxide Emissions from Field Crops. *Climate Change and Agriculture Fact Sheet Series,* Extension Bulletin E3152.

[38] Dessler, A., Sherwood, S., et al. (2009). A Matter of Humidity. *Science*, 323(5917),1020-1021. Retrieved from https:// science.sciencemag.org/content/323/5917/1020 and Dessler, A. and Schoeberl, M.R., et al. (2013) Stratospheric Water Vapor Feedback. Proceedings of the National Academy of Sciences, 110(45), 18307-18091. Retrieved from https://blog.pnas.org/ 2013/10/effect-of-stratospheric-water-on-climate/

[39] Malthus, T. (1798). *An Essay on the Principle of Population.* Retrieved from https://www.esp.org/books/ malthus/population/ malthus.pdf

[40] Union of Concerned Scientists. (1997). Retrieved from https:// www.ucsusa.org/sites/default/ files/attach/ 2017/11/ World% 20Scientists%27%20Warning%20to%20Humanity%201992.pdf

[41] Wynes, and Nicholas, K. (2017). The Climate Mitigation Gap: Education and Government Recommendations Miss the Most Effective

Individual Actions. *Environmental Research Letters,* 12(7). Retrieved from https:// doi.org/10.1088/1748-9326/aa754

[42] Ehrlich, P. R. and Holden, J. P. (1972). Critique, Bulletin of the Atomic Scientists. Vol. 28(5), Pgs. 16-27. Retrieved from https:// doi.org/10.1080/00963402.1972.11457930

[43] Meadows, D. H. et al. (1972). The Limits to Growth, a mass-marketed paperback.

[44] O'Neill, B.C. (2012*). Climate Change and Population Growth.* In Mazur, L. A. (ed.), 2012. *A Pivotal Moment: Population, Justice, and the Environmental Challenge.* Washington D.C.: Island Press. Also, O'Neill, B. and Bongaarts, J.(2018). Global Warming Policy: Is Population Left Out in the Cold? *Science,* 6403, 650-652. Retrieved from https://doi.org/10.1126/ Science.aat8680

[45] United Nations. (1992) Principle 8. *Rio Declaration on Environment and Development.* Report of the United Nations Conference on Environment and Development.

[46] O'Neill, B., et al. (2010). Global Demographic Trends and Future Carbon Emissions. *Proceedings of the National Academy of Sciences.* 107(41),17521-17526. Retrieved from https://www.pnas.org/ content/ pnas/107/41/ 17521.full.pdf

[47] Ibid.

[48] United Nations. Department of Economic and Social Affairs, Population Division. (2019*). World Population Prospects 2019: Data Booklet.* Retrieved from https://oi.org/ 10.18356/3e9d869f-en

[49] O'Neill, B., et al. (2010). Global Demographic Trends and Future Carbon Emissions. Proceedings of the National Academy of Sciences. Vol. 107(41) Pgs. 17521-17526. Retrieved from https://www.pnas.org/ content/pnas/107/ 41/17521.full.pdf

[50] WorldPopulationBalance.org

[51] O'Neill, B. et al. (2010). Global Demographic Trends and Future Carbon Emissions. *Proceedings of the National Academy of Sciences.* 107(41), 17521-17526. Retrieved from https:// www.pnas.org/ content/pnas/ 107/ 41/ 17521.full.pdf

[52] WorldWatchInstitute.org.

[53] Murtaugh, P. and Schlax, M. (2008). Reproduction and the Carbon Legacies of Individuals. *Global Environmental Change,* 19(1), 14-20. Retrieved from https://doi.org/10.1016/ j.gloenvcha.2008. 10.007

[54] Ibid.

[55] Wynes, and Nicholas, K. (2017). The Climate Mitigation Gap: Education and Government Recommendations Miss the Most Effective Individual Actions. *Environmental Research Letters,* 12(7). Retrieved from https:// doi.org/10.1088/1748-9326/aa754

[56] World Bank. Retrieved from https://data.worldbank.org/indicator/ EN.ATM.CO$_2$E.PC? locations=AU-US

[57] *Our World in Data,* based on the Global Carbon Project; Carbon Dioxide Information Analysis Centre using UN population estimates and 2020 Union of Concerned Scientists, International Energy Agency, Fuel Combustion Highlights.

[58] United Nations DESA. (2019) World Population Ageing Report 2019. New York, NY, United Nations.

[59] Ibid.

[60] World Health Organization. (no date). Under Five Mortality. *Global Health Observatory.* Retrieved from https://www.who.int/gho/child_ health/ mortality/mortality_under_five_text/en/

[61] United Nations. Department of Economic and Social Affairs, Population Division. (2019*). World Population Prospects 2019: Data Booklet.* Retrieved from https://oi.org/ 10.18356/3e9d869f-en

[62] Owen, D. (2010). Green Metropolis. New York, NY: Riverhead Books.

[63] Brown, M.A., Southworth, F., et al. (2008). *Shrinking the Carbon Footprint of Metropolitan America.* Washington, D.C.: Brookings Institution. Retrieved from https://www.brookings.edu/wp-content/ uploads/2016/ 06/carbonfootprint_report.pdf

[64] Ibid.

[65] Pradhan, E. and Canning, D. (2015). The Effect of Schooling on Teenage Fertility: Evidence from the 1994 Education Reform in Ethiopia. *PGDA Working Paper 12816.* Program on the Global Demography of Aging.

[66] UN Secretary General, Antonio Guterres, 2/2020speech at New School. Retrieved from https://www.un.org/sg/en/content/sg/speeches /2020-02-27/remarks-new-school-women-and-power

[67] WorldPopulationBalance.org.

[68] Wilson, E.O. (2003). *The Future of Life*. New York, NY Vintage and Wolchover, N. (2011) How Many People Can Earth Support. *LiveScience*. Retrieved from https://www.livescience.com/16493-people-planet-earth-support.html

[69] United Nations. The Food and Agriculture Organization. (2009). *How to Feed the World in 2050*. Retrieved from https://www.fao.org/ fileadmin /templates/wsfs/docs/ expert_paper/How_to_Feed_ the_ World_in_2050.pdf

[70] Retrieved from https://www.worldbank.org/en/topic/water-in-agriculture and WHO/UNICEF Joint Monitoring Programme for Water Supply, Sanitation and Hygiene.

[71] United Nations Water Division

[72] United Nations. The Food and Agriculture Organization, the International Fund for Agricultural Development and the World Food Program. (2013). *The State of Food Insecurity in the World*: *The Multiple Dimensions of Food Security*. 8. Retrieved from https:// www.fao.org/docrep/ 018/i3434e/ i3434e.pdf

[73] Steinfield, H., Gerber, P., et al. (2006). *Livestock's Long Shadow*. New York, NY: The Food and Agriculture Organization of the United Nations, p. xxii. Retrieved from https://www.fao.org/ docrep/ 018/1343e/i343e.pdf

[74] Hansen, M.C., Potapov, P.V, et al. (2013) High Resolution Global Maps of 21st Century Forest Cover Change, *Science*. Vol. 342, Pgs. 850-853.

[75] Steinfield, H., Gerber, P., et al. (2006). *Livestock's Long Shadow*. New York, NY: The Food and Agriculture Organization of the United Nations, p. xxii. Retrieved from https://www.fao.org/ docrep/018/ 1343e/i343e.pdf

[76] United States Department of Agriculture. (1995). Animal Manure Management Brief No. 7, Natural Resources Conservation Service. Washington, D.C.

[77] Steinfield, H., Gerber, P., et al. (2006). *Livestock's Long Shadow*. New York, NY: The Food and Agriculture Organization of the United Nations, p. xxii. Retrieved from https://www.fao.org/docrep/ 018/ 1343e/i343e.pdf

[78] Pinentel, D. and Burgess, M. (2013). Soil Erosion Threatens Food Production." *Agriculture*, 3(3), 443-463.

[79] Steinfield, H., Gerber, P., et al. (2006). *Livestock's Long Shadow*. New York, NY: The Food and Agriculture Organization of the United Nations, p. xxii. Retrieved from https://www.fao.org/ docrep/ 018/ 1343e/i343e.pdf

[80] Goodland, R. J. and Anhang, J. (2009). Livestock and Climate Change: What if the Key Actors in Climate Change were Pigs, Chickens, and Cows? *Worldwatch*, 22, 10–19. Retrieved from https://www.worldwatch.org/ files/pdf /Livestock% 20and% 20Climate%20Change.pdf 59.

[81] World Resources Institute and World Wildlife Organization

[82] Ibid.

[83] Zhang, H. and Schroder, J. (2014). Animal Manure Production and Utilization in the US. *Applied Manure and Nutrient Chemistry for Sustainable Agriculture and Environment*, 1-21. Stillwater, Oklahoma: Oklahoma State University. Retrieved from https:// doi.org/ 10.1007/978-94-017-8807-6.

[84] Steinfield, H., Gerber, P., et al. (2006). *Livestock's Long Shadow*. New York, NY: The Food and Agriculture Organization of the United Nations, p. xxii. Retrieved from https://www.fao.org/ docrep/ 018/1343e/i343e.pdf

[85] Alexander, P., Brown, C., et al. (2016). Human Appropriation of Land for Food: The Role of Diet. *Global Environmental Change*, 41. Retrieved from https:// www.research. ed.ac.uk/portal/files/ 28108199/ half_s3_v2.pdf

[86] Simon, D. (2013). *Meatonomics*. Newburyport, MA: Conari Press.

[87] Steinfield, H., Gerber, P., et al. (2006). *Livestock's Long Shadow*. New York, NY: The Food and Agriculture Organization of the United Nations, p. xxii. Retrieved from https://www.fao.org/docrep/ 018/1343e/i343e.pdf

[88] Simon, D. (2013). *Meatonomics.* Newburyport, MA: Conari Press

[89] Ibid.

[90] Williams, A. C. and Hill, L. J. (2017). Meat and Nicotinamide: A Causal Role in Human Evolution, History, and Demographics. *International Journal of Tryptophan Research,* 10, 1–23.

[91] Cassidy, E., West, P., et al. (2013). Redefining Agricultural Yields, from Tonnes to People Nourished per Hectare. *Environmental Research Letters*, 8(03). Retrieved from https://iopscience.iop.org/article/ 10.1088/1748-9326/8/3/034015/pdf

[92] World Wildlife Fund. Retrieved from https://www.worldwildlife.org/ threats/overfishing.

[93] Ibid.

[94] Ibid

[95] Ibid.

[96] Mongabay Newletter. October 30, 2008

[97] Retrieved from https://www.blueridgeaquaculture.com

[98] Gerten, D., Heck, V., et al.(2020). Feeding Ten Billion People is Possible Within Four Terrestrial Planetary Boundaries. *Nature Sustainability.*

[99] Campbell, T.C. and Campbell, T. M. (2006). *The China Study*, Dallas, TX: BenBella Book.

[100] Alexandratos, N. and Bruinsma, J. (2012). World Agriculture Towards 2030/2050: the 2012 Revision. ESA *Working Paper No. 12-03,* Rome, UN Food and Agriculture Organization

[101] McDowell, B.M., Hendricks, R.C., et al. (2011). *NASA's GreenLab Research Facility: A Guide for a Self-Sustainable Renewable Energy Ecosystem.* Cleveland, OH: National Aeronautics and Space Agency, Cleveland, OH. John Glenn Research Center. Retrieved from https:// ntrs.nasa.gov/archive/ nasa/casi.ntrs.nasa.gov /20120001794.pdf

[102] Radich, T. (2015). The Flight Paths for Biojet Fuel. *US Energy Information System Administration Working Papers Series.* Retrieved from https://www.eia.gov/workingpapers /pdf/flightpaths_biojetffuel .pdf

[103] Ibid.

[104] Coady, D., Parry, I., et al. (2019). Global Fossil Fuel Subsidies Remain Large: An Update Based on Country-Level Estimates. *International Monetary Fund Working Paper No. 19/89.* Washington, D.C. International Monetary Fund.

[105] Cornel, P. and Schaum, C. (2009). Phosphorus recovery from Wastewater: Needs, Technologies and Costs. *Water and Science Technology,* 59, 1069-1076.

[106] Ohtake, H. (2018) Phosphorus Recovery and Reuse from Wastewater. *International Water Association.* Retrieved from https://iwa-network.org/ phosphorus-recovery-and-reuse-from-wastewater/

[107] Kobayashi-Solomon, E. (2019). Investing in Vertical Farming: Five Take-Aways. *Forbes.* Retrieved from https://www.forbes.com/sites/ erikkobayashisolomon/2019/04/05/investing-in-vertical-farming-five-take-aways/#7e484946355c

[108] Varanasi, A. (2019). Does Carbon Capture Technology Really Work? *News from the Earth Institute.* Columbia University. Retrieved from https://blogs.ei.columbia.edu/2019/09/27/carbon-capture-technology

[109] Retrieved from https://www.climeworks.com

[110] Trent, J. (2012). *Offshore Membrane Enclosures for Growing Algae(OMEGA) - A Feasibility Study for Wastewater to Biofuels— Final Project Report.* Publication number: CEC-500-2013-143. Moffett Field, CA: NASA Ames Research Center.

[111] Schwartz, D. (2011). NASA's Omega Scientist, *Dr.* Jonathan Trent. *Algae Industry Magazine.* Retrieved from https:// www.algaeindustrymagazine.com/nasas-omega-scientist-dr-jonathan-trent/

[112] Trent, J., Kerekeš, K., et al. (2018). "UpCycle Systems: MPowered Farm Project Methods References Contact Information." Retrieved from https://doi.org/10.13140/ RG.2.2.28730.49607

[113] Trent, Jonathan, LinkedIn Interview on December 1, 2019.

[114] Wood, D. (2014). Space-Based Solar Power. *United States Department of Energy* Retrieved from https://www.energy.gov/articles/space-based-solar-power

[115] Space Solar Power: A New Beginning (2018). Retrieved from https://www.caltech.edu/about/news/space-solar-power-new-beginning-84155 and Solar Space Power Project, www.spacesolar.caltech.edu

[116] Coady, D., Parry, I., et al. (2019). Global Fossil Fuel Subsidies Remain Large: An Update Based on Country-Level Estimates. *International Monetary Fund Working Paper No. 19/89.* Washington, D.C.: International Monetary Fund.

[117] Kunkel, B. (2018). Ecologies of Scale. *New Left Review.* Retrieved from https://newleftreview.org/issues/II109/articles/herman-daly-benjamin-kunkel-ecologies-of-scale

[118] The World Bank. Wealth Accounting and Valuation of Ecosystem Services. Global Partnership Program: Frequently Asked Questions. Retrieved from https:// www.worldbank.org/en/news/feature /2015/06/15/ waves-faq

[119] Costanza, R. and Kubiszewski, I. (2014) Time to Leave GDP Behind. *Nature*, Vol. 505, Pgs. 284-6. Retrieved from https:// www.nature.com/news/polopoly_fs/1.14499!/menu/main/topColumns/topLeftColumn/pdf/505283a.pdf

[120] Costanza, R., DeGroot, R., et al. (1996). The Value of the World's Ecosystem Services and Natural Capital. *Nature,* 387, 253-260. Retrieved from https://www.researchgate.net/publication/40197297_

[121] UN FAO (2014) Food Wastage Footprint Full-Cost Accounting. Final Report. Retrieved from www.fao.org/3/a-i3991e.pdf

[122] Berik, G. and E. Gaddis. (2011). The Utah Genuine Progress Indicator (GPI), 1990 to 2007: A Report to the People of Utah. *Utah Population and Environment Council.* Retrieved from https://utahpopulation.org/wp-content/uploads/2014/11/ Utah_GPI_Report_v74 _withabstract.pdf

[123] Steffen, W., Richardson, K., Rockström, J., et al. (2015). Planetary Boundaries: Guiding Human Development on a Changing Planet. *Science,* 347, 622. https://science.sciencemag.org/ content/ 347/ 6223/ 1259855.full

[124] Ibid.

[125] Ibid.

[126] Ibid.

[127] Hickel, J. and Kallis, G. (2019). Is Green Growth Possible? *New Political Economy*, Retrieved from https://doi.org/10.1080/13563467.2019.1598964

[128] Dittrich, M., Giljum, S., et al. (2012). *Green economies around the world? Implications of resource use for development and the environment*. Sustainable Europe Research Institute. Retrieved from https://www.greengrowthknowledge.org/sites/default/files/downloads/resource/Green_economies_around_the_world_resource_use_SERI.pdf

[129] Schandl, H., et al. (2016). Decoupling Global Environment Pressure and Economic Growth: Scenarios for Energy Use, Materials Use and Carbon Emissions. *Journal of Cleaner Production,* 132, 45-56.

[130] Daly, H. E. (1990). Toward some Operational Principles of Sustainable Development. *Ecological Economics*, 2(1), 1–6.

[131] Hawken, P. ed. (2017). *Drawdown: The Most Comprehensive Plan Ever Proposed to Reverse Global Warming*. New York, NY: Penguin Books.

Important Note: The following is from the United Nations Department of Economic and Social Affairs -World Population Policies and generally applies to the contents of this book.

The regions are classified as belonging to either of the two general groups: more-developed and less-developed regions. The more-developed regions comprise all regions of Europe plus Northern America, Australia, New Zealand, and Japan. The terms 'more-developed regions' and 'developed regions' are used interchangeably. Countries in the more-developed regions are denominated 'developed countries.' The less-developed regions comprise all regions of Africa, Asia (excluding Japan), Latin America and the Caribbean, plus Melanesia, Micronesia and Polynesia. The terms 'less- developed regions' and 'developing regions' are used interchangeably. The designations 'developed' and 'developing countries,' 'developed' and 'developing' regions, and 'more-developed' and 'less- developed' regions are intended for convenience and do not necessarily express a judgment about the stage reached by a particular country or area in the development process.